Universe is a hologram

We live in a simulation

E. Van Herwi

WARNING

This book essentially offers theories scientists of the standard model and mechanics quantum, but also passages presenting a cosmological model not recognized by established science.

Willingly ignored for reasons of interest financial and geopolitical, this model explains why we are actually living inside a globe and not on the area.

This is called the theory of concave
Earth. (see *https://laterreestconcave.home.blog/*)

If I wanted to incorporate it into this work, it is because it can constitute major proof in principle holographic and the simulation hypothesis.

Summary

1 / Introduction

"Reality is only an illusion, although very persistent. "

Albert Einstein

While it constitutes one of the most completed and explains a lot of things, the model standard is far from perfect. Some theoretical physicists have now been working a few years on a new physics going to beyond the standard model, which could modify it by so that it agrees with the existing data.
There are physical phenomena in nature that the standard model can not explain:

1 / Gravity: The standard model does not explain it. A simply adding a graviton does not recreate what is observed experimentally without immediately requiring other changes still undiscovered in the model. At the quantum scale, gravity does not present no symmetry.

2 / Dark matter and dark energy: the cosmological observations indicate that the model standard explains only about 4% of the energy present in the universe. 27%, of the missing 96%, would be dark matter not interacting with fields of the Standard Model than through the weak interaction.
The rest of the 96% would be dark energy, a density constant energy for vacuum. Attempts explanation of dark energy in terms of the energy of the vacuum of the standard model lead to a difference in 120 orders of magnitude.

3 / The mass of neutrinos: according to the standard model, neutrinos are massless particles. The quantum physics with experiments on their oscillation has shown that they have one. However, although they are the second most abundant in the universe after photons, these particles are elusive. They never

interact with matter, which they cross at a speed close to of that
of light. They are only energy and are electronvolt measurement.

4 / Matter-antimatter asymmetry: The model
standard predicts that matter and antimatter were
created in almost equal quantities because the initials conditions
of the universe did not imply a disproportion
between one and the other. Yet no mechanism does not appear
sufficient in the standard model for explain this asymmetry.
You should also know that our world has a number of physical
characteristics enough confusing, such as:

- Quantum superposition
- The quantum entanglement
- The wave-particle duality
- Planck length
- The relativity of time

There are also philosophical characteristics disconcerting:

- The problem of personal identity
- The problem of the passage of time
- The problem of free will

Quantum physics brought about a revolution
conceptual in philosophy and literature but it has
especially allowed to develop many technological applications
such as energy nuclear, medical imaging, transistor, circuit
integrated, the electron microscope, the laser ...
A century after its conception, it is widely used in
research in theoretical chemistry (quantum chemistry), in physics
(quantum mechanics, quantum theory of fields, condensed
matter physics, physics nuclear, particle physics, physics
quantum statistics, astrophysics, quantum gravity), in
mathematics (formalization of the theory of fields) and, recently,
in computer science (computer quantum, quantum

cryptography). More than anything, quantum physics today allows analyze the reality of the world around us. But what is reality apart from energy and information?

2 / The big bang; a pulse

2.1 / The universe began with an impulse

For over a hundred years, scientists have postulated
that the universe had started with a Big Bang, that all
the universe was compressed into an infinitely small point and
that in a fraction of a second he began to develop outward. But if
the universe had not started with an impulse rather than an
explosion? This idea was supported by Dr Abhay Ashtekar, from
Penn State University. There's a lot inconsistencies between the
cosmic background of microwaves and our current model of
cosmology:

1.On a very large scale, the universe does not act as it should
thought.
2.The material acts like a giant lens, folding and changing the
amplitude of the light behind it, which is not consistent with our
current model of the universe.

3.The two hemispheres of the sky CMB (dark matter
cold) have different average temperatures so that scientists think
the universe should have start with a uniform temperature on
average.
4.The value of the Hubble constant, which describes the speed
expansion of the universe, is different if we measure from CMB
or Cepheid stars plus relatives.

All of these anomalies mean we're missing something
fundamental in our understanding of the universe.
The key could lie in quantum cosmology in loop (LQC).
LQC comes from looping quantum gravity which
is made up of particles called quanta. These quanta
form the fabric of space and time.
In this model of the universe there is the smallest size of the
space itself - the Planck scale - i.e. 10^{3-5} meters. That is to
say :
0.0000000000000000000000000000000001 meter!
Nothing can be smaller than this. It means the Big Bang couldn't
exist in a universe with LQC. The universe could never come
down to a point infinitely small and infinitely dense. Close to the
Big Bang, when the universe was very small, things really bizarre
arises mathematically. The infinites began to pop up and
threaten to tear the fabric of space-time itself. It is in these places
that the loop quantum gravity can intervene to make corrections
to classical physics.
Quantum corrected equations predict that there is
an effective repulsive force. This means that in these
regions where the universe is very small, the repulsion would
cause rebound. If our Universe was born from a rebound, there
was then another universe before ours.

2.2 / Fine tuning of the universe

Setting	Max. deviation
Ratio of electrons: Protons	1:10 37
Electromagnetic force ratio: gravity	1:10 40
Expansion rate of the universe	1:10 55
Density of the universe	1:10 59
Cosmological constant	1:10 120

These numbers represent the maximum deviation from accepted values, which would not be suitable for any form of life.

Recent studies have confirmed the fine tuning of the cosmological constant, also called black energy. This cosmological constant is a force which increases with the increasing size of the universe. The recent measurements of the cosmological background of micro-waves (CMB) do not only demonstrate the existence of the cosmological constant, but also its value. He turns out that the latter exactly compensates for the lack of matter in the universe.

The ripples in the universe of the original event of the Big Bangs are detectable in one part in 100,000. If this factor was slightly smaller, the universe would not exist as a collection of gases. If this factor were slightly larger, the universe would be made up only large black holes.

Another finely tuned constant is the force strong nuclear power (the force that holds atoms together). The Sun fusing hydrogen and higher elements together. When both hydrogen atoms merge, 0.7% of the mass of hydrogen is converted into energy. If the amount of converted material was slightly less, 0.6% than instead of 0.7%, a proton could not bind to a neutron, and the universe would only be of hydrogen. If the amount of material converted was slightly higher, 0.8%, the merger would occur if easily and quickly that no hydrogen would have survived the Big Bang.

Precise fine-tuning parameters for the universe:

1.Steady strong nuclear force
2.Weak nuclear force constant
3.Constant gravitational force
4.Electromagnetic force constant
5.Report of the force constant
electromagnetic at the gravitational constant force
6. Ratio of the electron to the mass of the proton
7 Ratio of the number of protons to the number
of electrons
8 level of entropy of the universe
9.Mass density of the universe
10.Speed of light
11.Age of the universe
12. Initial uniformity of radiation
13.Average distance between galaxies
14.Density of the galaxy cluster
15.Average distance between stars
16 Fine structure constant (describing the division
fine structure of the spectral lines)
17. Proton decay rate
18.Nuclear power level ratio of 12 C at 16 O
19. Ground state energy level for 4 He
20. Decay rate of 8 Be
21.Report of mass of neutrons to mass proton
22.Initial excess of nucleons compared to anti-nucleons
23.Polarity of the water molecule
24.Supernova eruptions
25.White Dwarf Binaries
26 Ratio of the mass of exotic matter to the mass of ordinary
matter
27.Number of effective dimensions in the primitive universe
28. Number of effective dimensions in the current universe
29. Neutrino mass

30.The big bang waves
31.Size of the relativistic expansion factor
32. Magnitude of the uncertainty in the principle
of Heisenberg uncertainty
33.Cosmological constant

3 / Abiogenesis is impossible

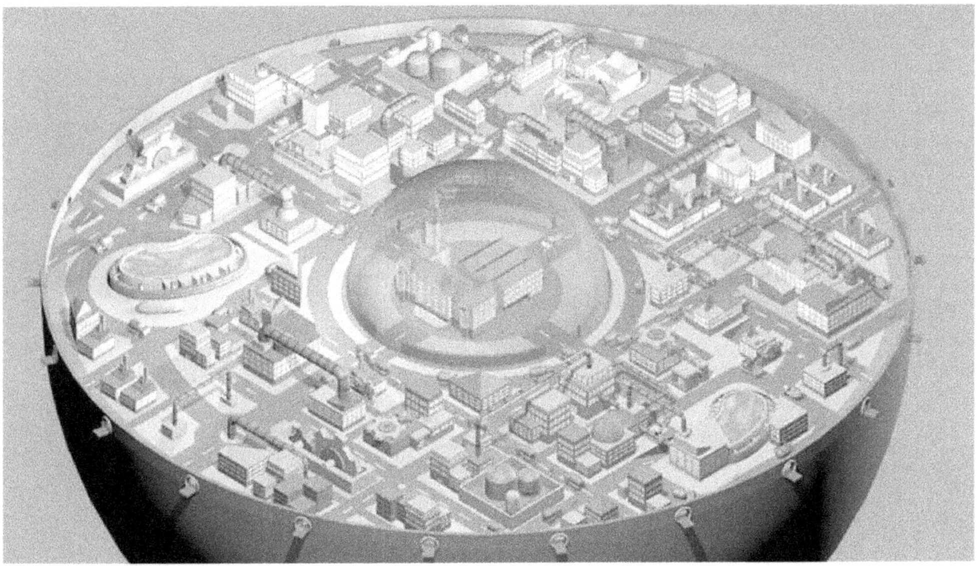

The origin of life depends on the biological cells that perpetuate life through the complex action of:

- The cell membrane or plasma membrane is composed of phospholipids and proteins. She is permeable and elastic and envelops the cell, the delimits and controls cellular exchanges.
- Organelles
• Mitochondria provide energy to the cell and ensure cellular respiration.
• Ribosomes make proteins.
• The granular endoplasmic reticulum or rough receives the proteins made by ribosomes and sends them to the Golgi apparatus.
• The Golgi apparatus stores and distributes protein.
• Lysosomes destroy the elements undesirable to the cell by digestion enzymatic.
- Chromosomes and gene regulation network

- DNA
- RNA polymerase

PLANT CELL

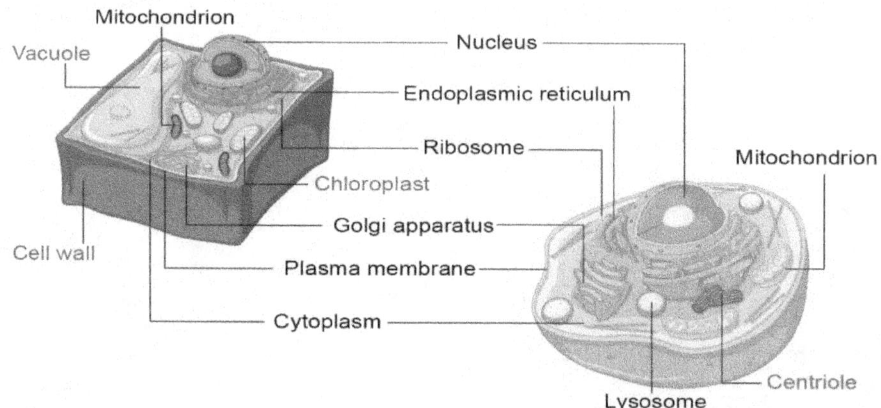

Human and plant cells have in common:

- The nucleolus
- The core
- The ribosome
- Smooth endoplasmic reticulum
- Mitochondria
- The peroxisome
- The plasma membrane
- The golgi apparatus

Francesco Redi

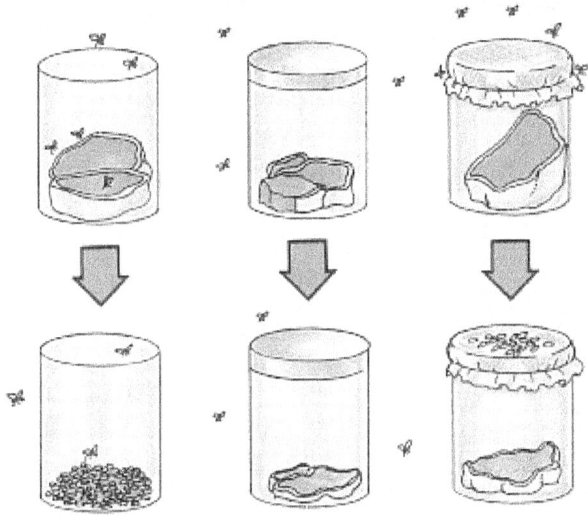

We know today that life only comes from life, that is, living beings come from beings pre existing. Absolutely everything was created.

The belief in spontaneous generation took a long time part of common sense to explain the appearance of beings alive where they could not be seen. little mice could be born spontaneously from maggots or take out a piece of meat. As for micro-organisms, it seemed that bacteria and yeast were the obvious product of spontaneous generation.

To verify this assertion, Francesco Redi made an experience. Francesco Redi, Italian scientist born in 1626, inspired by the work of William Harvey, published `` *Experiments on the generation of insects* " in 1668.

This work provided evidence against the theory of spontaneous generation which says that living beings can form from non-living objects. In experiment, Redi prepared three groups of jars, each with pieces of meat inside. A group of jars was covered with gauze, one group was left open and one group was completely sealed.

In the group of jars left open, Redi found larvae on meat. Redi noticed that in jars that were completely sealed, there was no maggot. In the group of jars that were covered with gauze, he noticed that there was no maggot on the meat, but larvae appeared on the gauze. This experiment provided evidence that refuted the theory of spontaneous generation. He showed that the spirits came from eggs laid by flies. This experience was important because it was one of the first controlled in history.

Pasteur

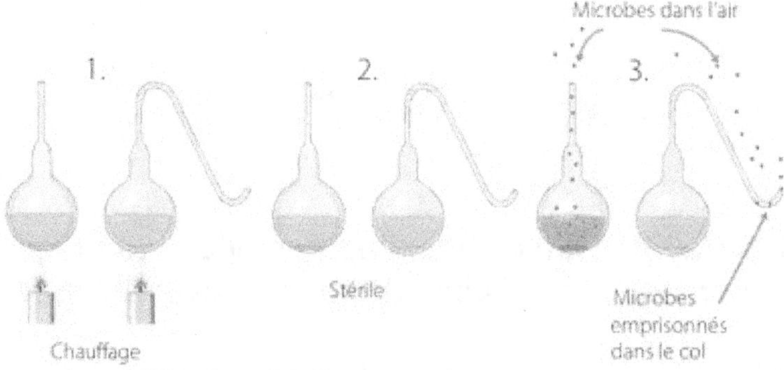

In 1864, the chemical biologist Louis Pasteur undertook to put an end to the postulates of spontaneous generation. For to achieve this goal, Pasteur produced containers in glass called gooseneck flasks, then boiled a series of broths which remained sterile. When the neck of one of them broke, it became contaminated and the micro-organisms proliferated rapidly. The evidence provided by Pasteur were irrefutable, succeeding in destroy a theory that has persisted for more than 2,500 years.

4 / The atom is 99% empty

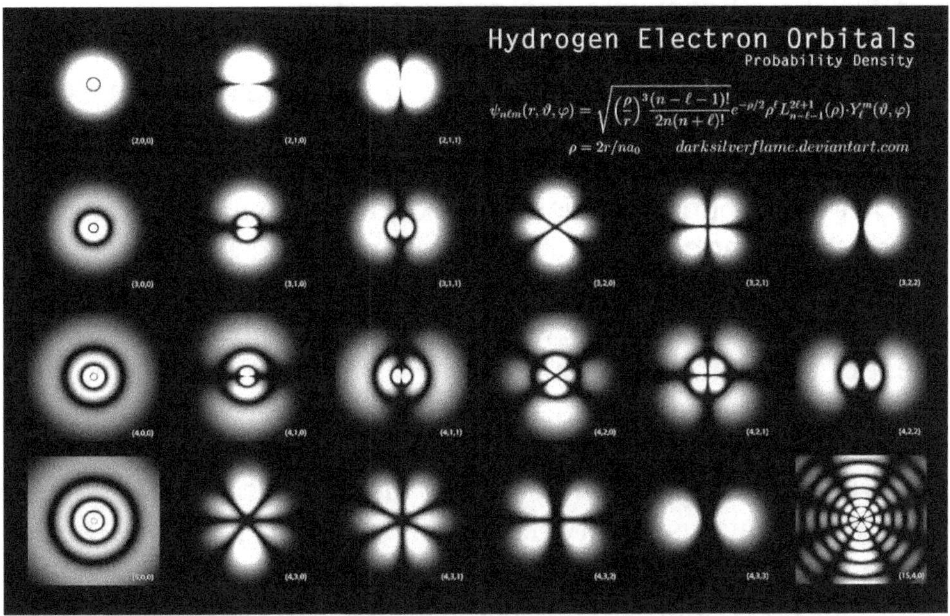

*Typical behavior of electron density in an atom
for different electronic orbitals.*

Contrary to what we may have thought for centuries, our world is not made of things solid. Since our world is made up of atoms without physical mass but only energy, this therefore means that the ground on which we walk is 99% empty.
Physicists now describe matter by its most basic level like clouds of probabilities or quantum foam. The principle of uncertainty of Heisenberg says that one can never know the precise location of particles while knowing where they are going. Their very existence is a mix between a wave of potential and an entity physical. It's a very strange world when you are there take a closer look. Quanta in quantum physics consist of discrete amounts of energies or of states in which a particle can exist. The Newton's equations assumed a continuous quantity

space; it turns out that the universe is maybe more quantified than we thought.

So if the atom is 99% empty why is a table does it feel so solid? Well it's because of the dance electrons!

If you touch the table, the electrons of the atoms of your fingers approach the electrons of the atoms of the table. As the electrons of an atom collect sufficiently close to the nucleus of the other, the patterns of their dances change. A weak electron energy level around a nucleus cannot make up the same thing around the other because this slot is already occupied by one of its own electrons. New come must take an unoccupied role and more energetic. This energy must be provided, not by light this time, but by the force of your finger which probe. So, already growing two atoms near one of the other takes energy, because all their electrons must enter high energy states unoccupied. Trying to bring together all the atoms in the table and all finger atoms is very demanding of energy. We feel it, like resistance on our finger, that's why the table is solid to the touch. Inside our body, there is mainly a series clouds of electrons, all bound by quantum rules that govern the entire universe.

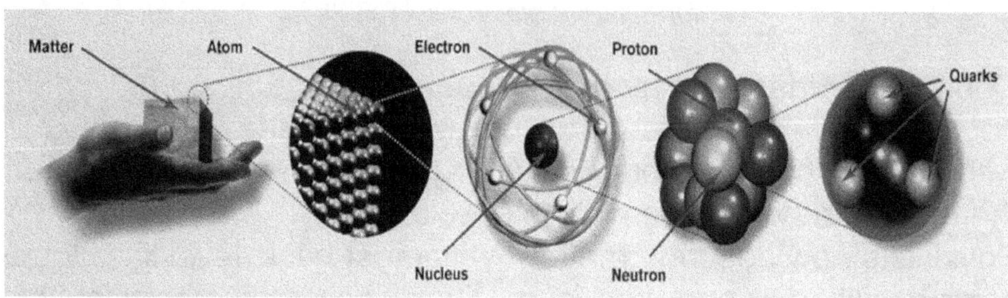

The size of the nucleus of an atom is of the order of 10-15 m, while the atom is about 10- 10 m in diameter.

If the nucleus was 1 cm in diameter, the electron would thus gravitate in a volume of about 1 km of diameter!

Light made up of electromagnetic waves propagates in a vacuum; it should therefore logically cross without hindrance the matter which should then seem transparent, but if you can see the objects well, is that the nucleus emits powerful fields electromagnetic systems that hold electrons in atom and ensure the cohesion of matter.

When light hits an atom, it puts in vibration the electrons which disturb the field electrical and thus modifies the radiation received. This is this modification which makes that we see the material which is that made up of atoms. The latter does not absorb light at certain frequencies, called resonance, which depend on the level of its energy.

This is what gives color to objects. When the energy of the photons is not large enough to excite electrons, light waves pass through material without causing interference, and appear then transparent. By changing the wavelength of the light, it is possible to see through objects opaque. Low wavelength rays, such as x-rays make it possible to see through a body or detect objects in luggage.

5 / Nothing goes faster than the speed of light

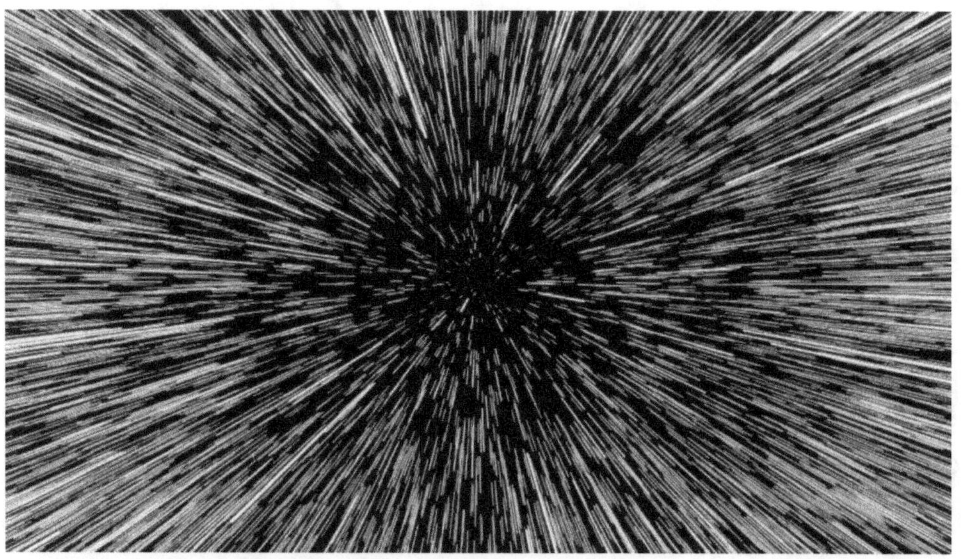

If the physical world is a virtual reality, it is the
information processing product. This is defined as a choice
among a finite set, so that the treatment which changes it must
also be finished, and in indeed, our world is cooling off at a finite
rate. A supercomputer processor refreshes 10 quadrillion times
per second, and our universe refreshes a thousand billion, trillion
times faster than that, but the principle is the same. As a screen
image has pixels and a refresh rate, our world has Planck length
and Planck time.
In this scenario, the speed of light is the fastest speed.
faster because the network cannot transmit anything anymore
faster than one pixel per cycle, that is, the length of Planck
divided by Planck's time, i.e. About 300 000 kilometers per
second. Relativity is a concept perfectly intuitive. In theory, it
states that any benchmark inertial is equivalent to another. An
inertial benchmark simply designates a point that you choose

as a landmark.

$$Planck\ length = \sqrt{\frac{hG}{c^3}} = 16229 \times 10^{-35} m$$

$$Planck\ mass = \sqrt{\frac{hc}{G}} = 2.17647 \times 10^{-8} kg$$

$$Planck\ time = \sqrt{\frac{hG}{c^5}} = 5.39116 \times 10^{-44} s$$

To be inertial, this reference point must be at constant speed and not in acceleration or deceleration. For example, suppose you were running at 20 km / h in a TGV traveling at 300 km / h. A observer on the platform of a crossing station would see you pass at 320 km / h, its inertial reference point is the platform of the station. Yet for train passengers, whose inertial benchmark is the train itself, you are doing well 20 km / h. If the train rolls along the highway and crosses a car traveling at 120 km / h, the driver of the car would see you go to 440 km / h, its inertial benchmark being his car!
This relativity of velocities is called relativity Galilean. We experience it every day. Yes a flight attendant serves you coffee on an airplane flying at 800 km / h, the coffee falls straight into your cup and is not plastered on your shirt. The reason is that you, the mug, the hostess and the coffee all have the same speed. Relatively speaking, you are at 0 km / h one by in relation to each other, you are all in the same benchmark inertial while going at 800 km / h in relation to the ground which is another inertial benchmark.

Electromagnetism

In the 19th century, Maxwell, worked on the airwaves electromagnetic waves that allow transmit information remotely.

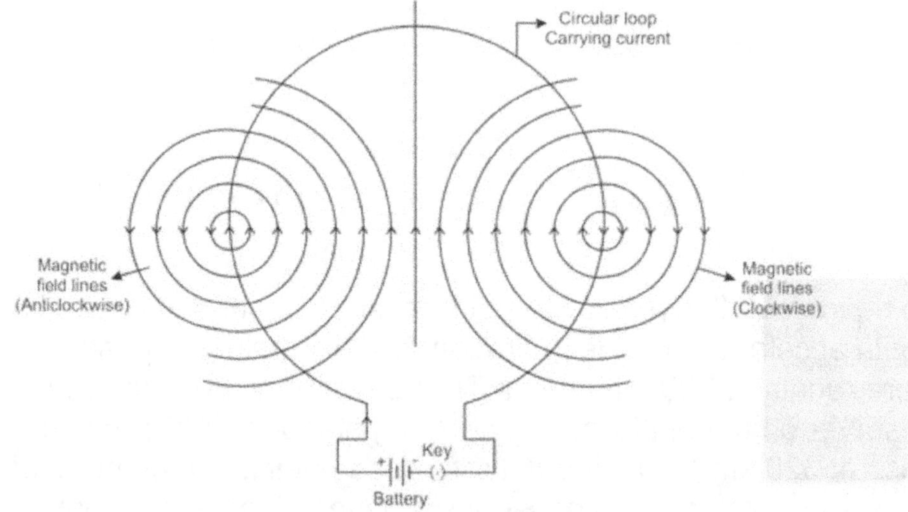

Magnetic field lines due to circular loop carrying current

Thanks to a complex mathematical work, he demonstrates that the speed of electromagnetic waves in a medium depends only on two coefficients: the permittivity electrical and magnetic permeability. For the void, these two values are known. The result of his equation is that the speed of a wave electromagnetic is extremely close to the value known speed of light. This proves without the shadow of a doubt what some suspected: the light is only a manifestation of waves electromagnetic. According to Maxwell, the speed of light depends only on the medium, not on the speed relative of this medium. The speed of light does not depend on not of the inertial frame. Whatever your speed by compared to the light source, you see the light at the same speed.

Ether

Ether would be a very thin material that would compose the whole universe.
Michelson and Morley attempt an experiment aimed at prove the existence of ether. To simplify, let's say that they will send two light rays traveling the same distance but perpendicular to each other. Indeed, if there is ether, the movement of the Earth in this ether must cause the equivalent of a breath. Imagine on the roof of the TGV; you throw two marbles perpendicular to the roof, the wind caused by the speed will change their trajectory and you could determine the speed of the TGV or at least its direction. AT to everyone's surprise, the result of this experiment is without appeal: there is and cannot be ether. It is necessary sometimes a flash of genius to unblock a situation, and Einstein has it. He decides to apply " *we said that the speed of light is the*

same in all inertial frames of reference " and therefore submits this postulate to the place of the principle of simultaneity. The link between all the inertial reference points will now be the same and unique speed of light. Einstein's genius was therefore to discover that the passage of time actually depended on the speed at which we move. What would be this fundamental natural constant which defines the passage of time? The answer is clear: there is no not. The only fundamental constant is the speed of the light. The human always remains very close to one inertial reference point: the Earth. Einstein concludes on the ether by the following summary:

According to the theory of general relativity, space is endowed with physical properties, therefore, there are an ether. A space without ether is unthinkable, because in such a space not only would there be no light propagation, but also no possibility of existence for a standard space and time, nor for spacetime intervals in the physical sense of the term. However, this ether cannot be designed as provided with the qualities of ponderable media and as made up of parts having a trajectory in the time. The idea of movement cannot be him applied."

6 / The Goldilocks area

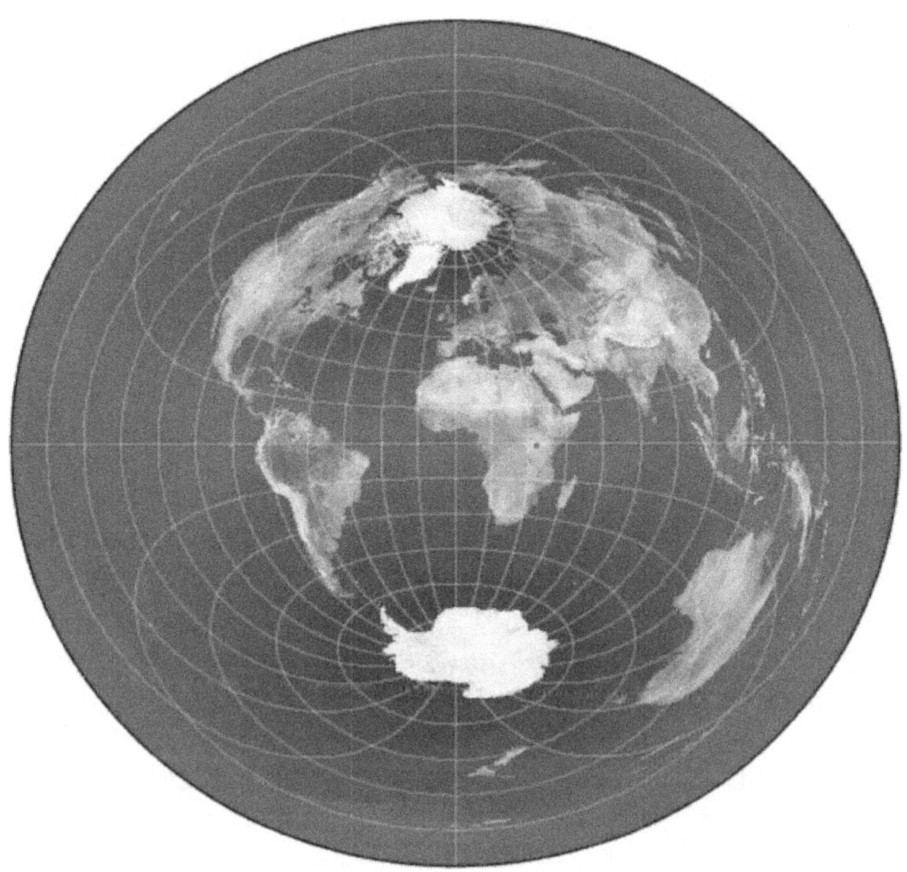

6.1 / The true shape of the Earth

In established science, the Goldilocks Zone is
reference to a habitable zone around a star where the
temperature is neither too hot nor too cold for liquid water is
found on a planet.
Liquid water is essential to life as we know it.

But we never found life on others planets, even in our own solar system. A abbreviated list of requirements for a suitable universe for life of any type should include the following:
• Elemental chemical stability and diversity
sufficient to build the molecules complexes necessary for vital functions essential: energy processing, storage
information and replication. A compound universe only hydrogen and helium do not will not work.
• Predictability of chemical reactions, allowing compounds to form from the different elements.
• A universal connector, an essential element for the molecules of life. He must have the property chemical that allows it to react easily with almost all other elements, forming stable bonds, but not too stable, so that disassembly is also possible. The carbon is the only element in the periodic table which meets this requirement.
• A universal solvent in which the chemistry of life can
unfold. Since the reactions chemicals are too slow in the solid state and the complex life would probably not be maintained in the form of gas, there is a need a liquid element or a compound that dissolves easily both reagents and products reaction essential to living systems: namely, a liquid with the properties of water.
• A stable source of energy to support the living systems in which there must be photons from the sun with sufficient energy to lead to organic and chemical reactions, but not energetic enough to destroy the organic molecules (as in the case of highly energetic ultraviolet radiation).
• A means of transporting energy from the source
(like our sun) to the place where occur chemical reactions in the solvent (such as water on Earth) must be available. In the process, there must be minimal losses in transmission if energy is to be used effectively.
The Kepler space telescope of the planet-hunting NASA searches for planets orbiting the areas habitable of stars similar

to the Sun by searching planets with an average orbit of 365 days.

It is not because a planet or a moon is found in the Goldilocks zone of a star that it will have life or even liquid water. Earth is not the only planet in the Loop gold zone. Venus and Mars are also in it, but not are not habitable. And for good reason these are only rocks of which the nature, nor the matter.

In fact, just as man has never been able to walk on the moon, we can never set foot on Mars. The Earth is very round, it is spherical but we are not on its surface. We live in the Earth. This is not the Hollow Earth theory but that of the concave Earth.

The Hollow Earth; a fake model

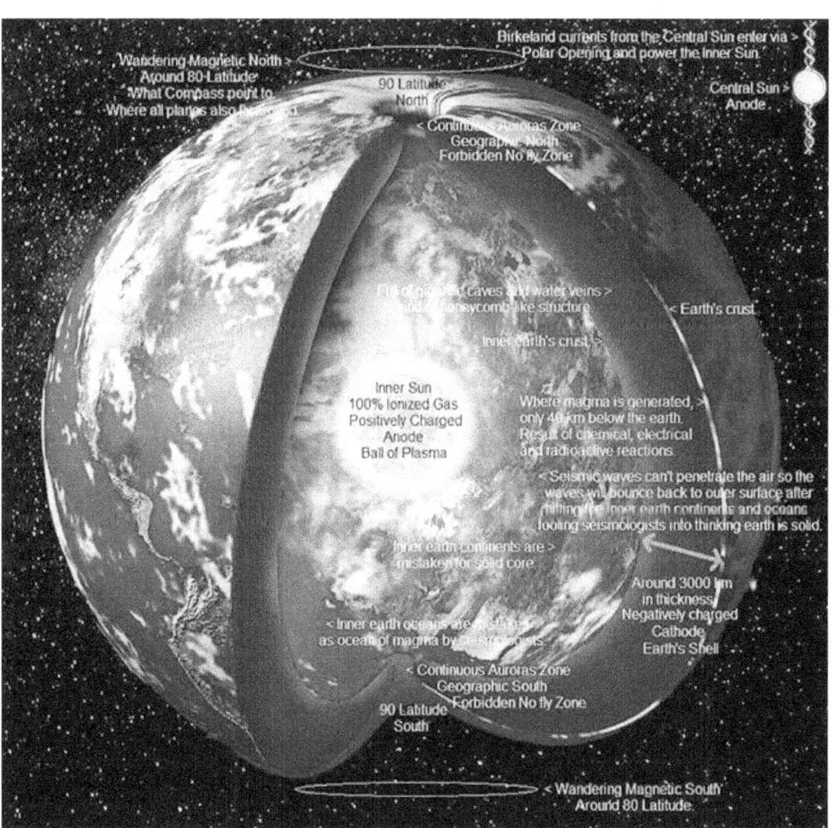

Edmund Halley put forward in 1692 the idea that the Earth was made up of a hollow shell of about 800 km of thickness, of two inner concentric shells, and a central core, having respectively the diameters approximations of the planets Venus, Mars and Mercury. These shells would be separated by a layer atmospheric, each of them would have their own poles magnetic, and would rotate at different speeds.
Halley proposed this model to explain anomalies in the display of the compasses. He emitted the hypothesis of the existence of a luminous atmosphere at the interior of the Earth, this one producing the aurora boreal by escaping outside. He also issued the hypothesis that the inner worlds could be inhabited. Some have claimed that Leonhard Euler had also put forward the idea of a Hollow Earth, eliminating the multiple shells to postulate the existence of a sun interior that would provide light to a civilization advanced. This design could come from the misinterpretation of a writing in which Euler related a simple thought experiment. Sir john Leslie later developed this idea, suggesting two central suns, which he named Pluto and Proserpina.
Later, in 1818, John Cleves Symmes, Jr. Suggested that the Earth consisted of a shell of about 1,300 km thick, with openings of about 2,300 km at the two poles, and four inner shells, each of which is also open to the poles. Symmes became the most famous among the first partisans of the hollow Earth. He prepared even an expedition to the North Pole, but the new President of the United States, Andrew Jackson, ended the attempt because he was convinced that the Earth was flat! Symmes died in 1829 without having been able to complete his project. One of his disciples, Jeremiah N. Reynolds, who organized conferences on the Hollow Earth, suggested also an expedition to the pole. It seems he tried organize one by himself, but the outcome remains obscure. Some people claim that it had only pecuniary interest, only the shipment that he was in fact only an attempt at fraud, and that he then disappeared. For others, he actually tried to complete his expedition but failed, then attempted vain to join the expedition of Charles Wilkes in 1838-

1842, the rest of his life being unknown to us. Symmes even wrote no books about his ideas, but others did it. McBride wrote The Theory of Spheres concentric lines of Symmes in 1826. It seems that Reynolds wrote an article which appeared as separate brochure in 1827: Commentaries on the theory by Symmes in the American Quarterly Review.

In 1868, a professor by the name of WF Lyons presented in The Hollow Globe a theory close to that of Symmes, but did not mention the latter. More recently an early Hollow Earth supporter 20th century William Reed wrote *The Phantom of the poles* in 1906. He proposed the idea of a Hollow Earth, but without shells or inner Suns. Then came Marshall Gardner who wrote *A Journey Into the Land* in 1913, then an enriched edition in 1920. He placed an inner sun in his Hollow Earth. He then builds some a functional model that he patented (#1096102). Gardner did not mention Reed, but took Symmes as source of his ideas. In 2001, Kevin and Matthew Taylor, father and son, published the book *La Terre sans horizon*, in which they propose a theory at least original in which the Earth is hollow, and in a phase expansion which must lead to a final state of equilibrium.

In their theory, the presence of a central sun of small size, fed by radiation from the surface interior of the Earth's shell, explains in particular terrestrial magnetism. According to mathematical theory of the gravitational potential of Isaac Newton, the gravitational force is zero inside a shell spherical, whatever its thickness, if we neglects the effect of other masses inside and outside the outside of the shell (so-called shell theorem dig). Thus, according to this theorem, the beings who would live in the interior of a supposed hollow Earth would suffer no attraction to the outside, and therefore could not stay on the ground. They would be in a state almost complete weightlessness, feeling only the slight residual force of gravity from the shape imperfectly spherical of the Earth, and the forces of tides produced by outer celestial bodies, like the moon. The centrifugal force due to the rotation of the Earth would in theory attract them outwards, but it would not exceed, even at the

equator, 0.3% of the force of gravity exerted on the "outer" surface of the Earth.
But it was without counting that in the reality of the concave Earth.

The real model; the concave Earth

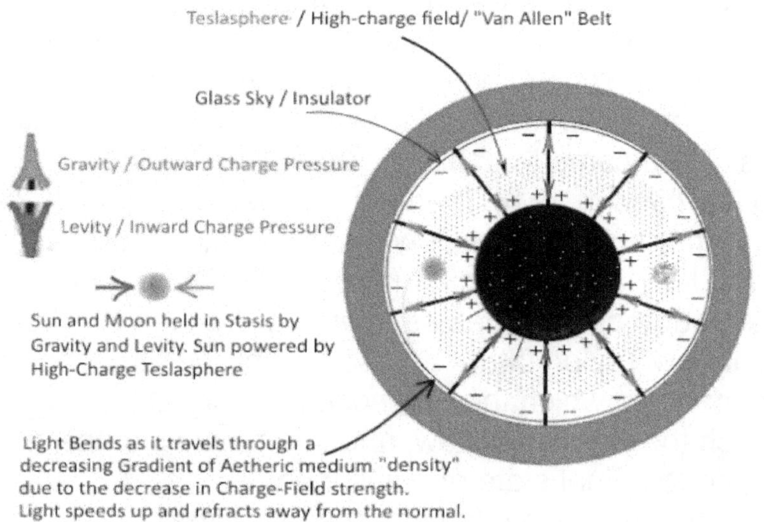

Instead of considering that we live on the surface exterior of a hollow planet, some theorists are supporters of a concave Earth theory. According to one of these theories we live inside a hollow world in which it is the centrifugal force and not the gravity that keeps us on the ground, and the universe that we let's see is only an illusion that could be produced by deviations of light. Cyrus Teed, doctor and alchemist, had in 1869 the mystical intuition of such concave Earth model, which he called *Cellular Cosmogony* . From this worldview, he founded the movement Koreshan Unity because Koresh is the version Hebrew name of his first name Cyrus, and in 1894 created a utopian community in Estero, Florida, today historical park.

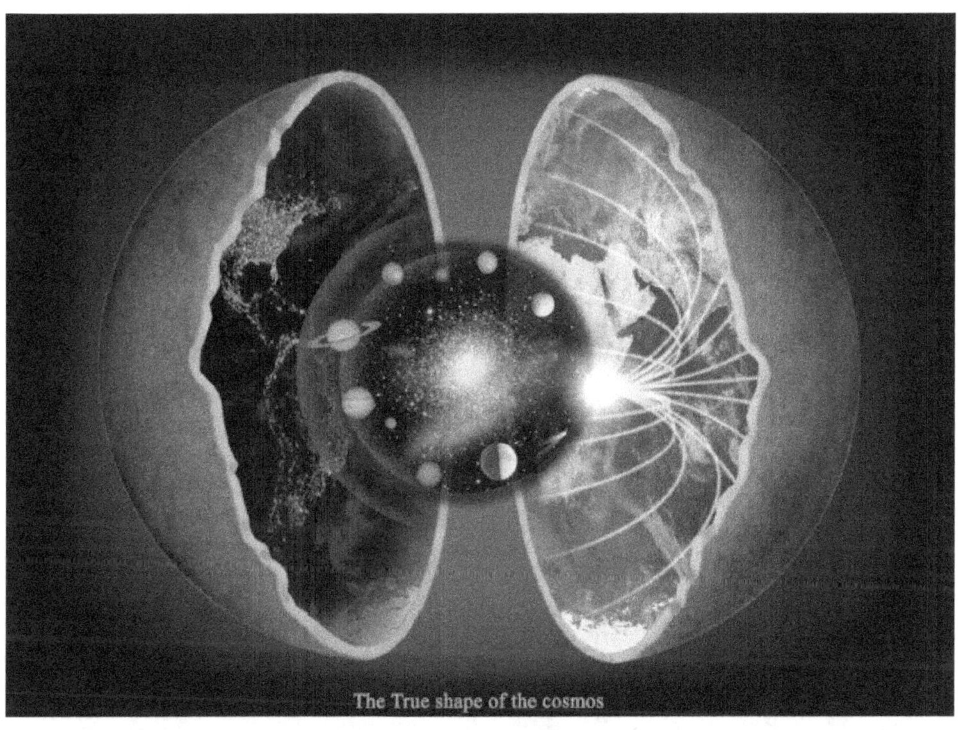

The celestial sphere is at the heart of the Earth

To confirm his enlightenment, Teed and his followers carried out
measurements on the Florida coast, near the city of Naples,
using technical equipment appropriate says the rectilineator; The
obtained results provided experimental proof of the concavity of
the curvature of the Earth. They were unfortunately not
taken seriously and was criticized by Donald E. Simanek.
Then several German writers of the 20th century, including Peter
Bender, Johannes Lang, Karl Neupert and Fritz Braun, published
works defending the theory of Concave Earth

(Hohlweltlehre). Several members of the entourage of Adolf Hitler, influenced by this ideology, allegedly ordered an operation to spy on the British fleet from the island of Rügen at sea Baltic; it was a question of obtaining images of the forces enemies by directing telescopes to the sky, which ended in failure. Today we (the concavists), know that it is impossible to see beyond of the horizon due to the successive layers of air which block vision beyond the curvature.

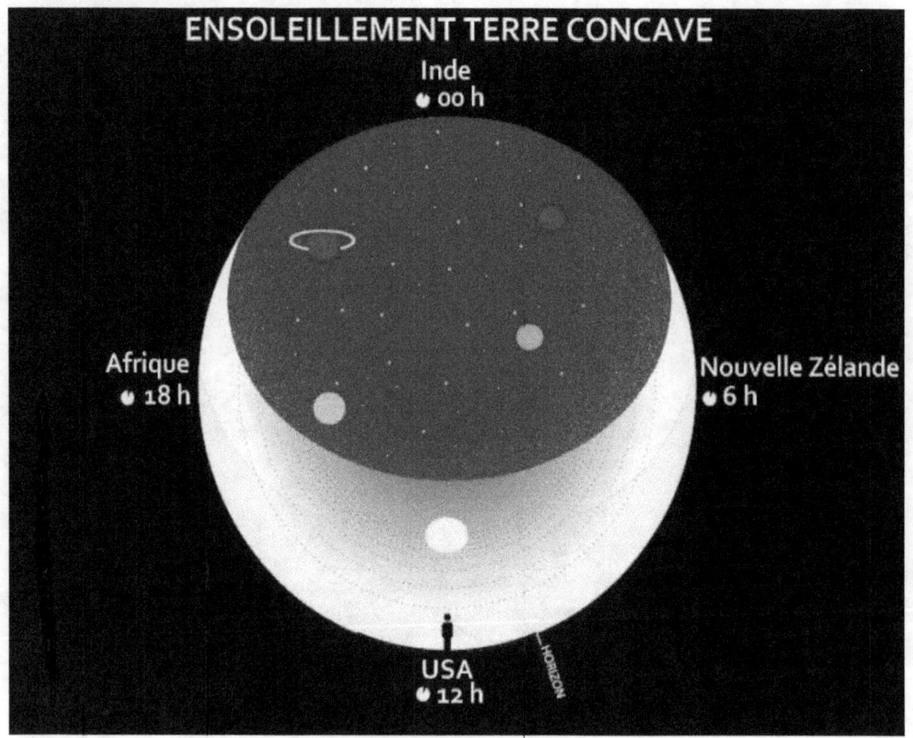

Sunshine in the Earth

In 1981, Mostafa Abdelkader, Egyptian mathematician of Alexandria, took over and developed the Geocosmos version by Karl E Neupert of Ideas by Cyrus Teed, dating from 1900. Unlike Teed's model which considers celestial bodies as optical illusions, the Neupert's model inverts all of the known cosmos. In the concave pattern, indicating that the space is narrowing / implodes via a non-Euclidean geometry, so as to place an entire

Copernican cosmos (C) in the enveloppe relatively finite limit of the concave surface from Geocosmos (G). In his document which he submitted to the Australian scientific journal; *Speculations in Science and Technology*, Abdelkader says:

Huge galaxies and other distant objects are mapped inside as objects microscopic, our moon being by far the largest celestial objects, which revolve around the axis every day of the Earth. The straight rays of light are represented as arcs of circles, so that all celestial phenomena appear inside observers in G as well as observers exteriors in C. We then consider the hypothesis according to which, conversely, our current universe is this G.finished. This idea involves the inversion of the whole known and established geophysics and astrophysics. "
The main addition to the concept of Neupert that Abdelkader has discussed, is that the light is ultimately attracted to the center of the cosmos which narrows inward. Bows of light that travel to the Earth's surface are absorbed, and those who are not continue to travel to the center of the cosmos and around it to the opposite side of the heavens. However, they never light up the other side of the Earth or its night sky because the wavelengths of light flow according to the volume of space beyond the surface of the Earth and undergo the inertia of the infinitesimal center.
So while they converge to the opposite position from the heavens to the place where the sun is, they are simultaneously attracted to the center.
Therefore, the light that revolves around the side opposite of the heavens, never meet the sight of those who have the center of the cosmos between them and the sun. A surface observer will experience the night without bright sky, even if the rays of light pass through really the space in which they are located, because the rays are only in space and are never received directly into the eye.
The mathematician devised an equation which suggests that the phenomena belonging to the perception of the sky-dome and

skyline are subject to illusion optics, supporting his claim that light is bends in wide arcs across space rather than travel in a straight line. As such, when we believe we are looking at an astrological object luminous perpendicular to our line of sight, which we see is actually at the other end of a bent ray of light.

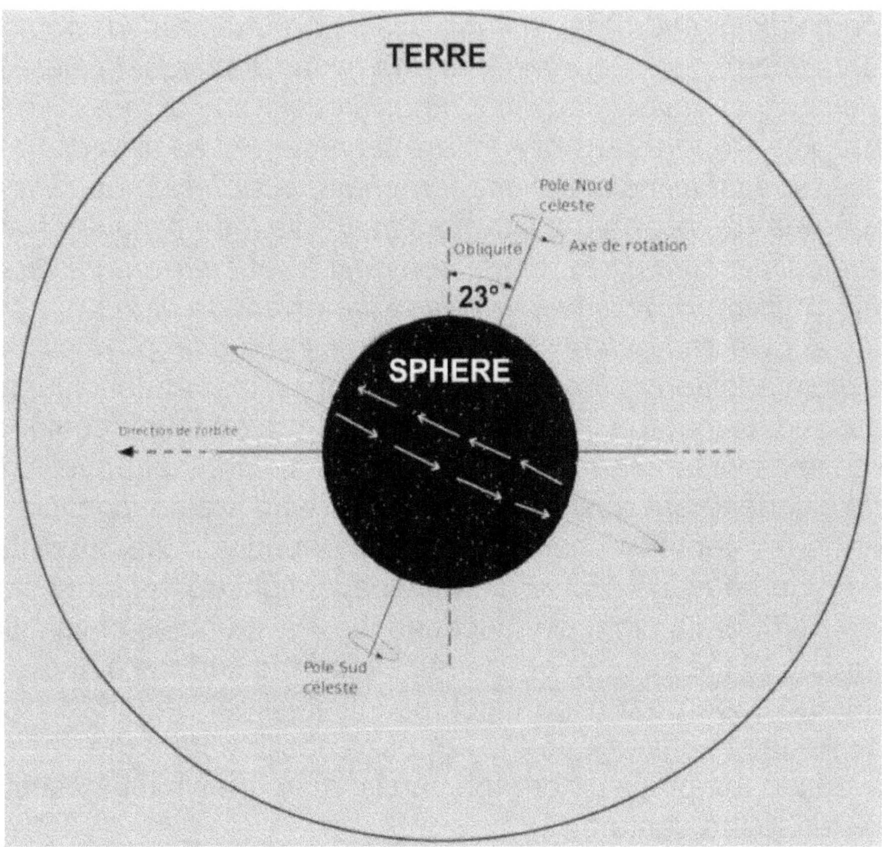

It is not the Earth which rotates but the celestial sphere which is inclined at 23 °

The line of sight isn't really pointing at anything, and their eyes receive an image of something that is in the sky, elsewhere, in the direction they are looking really. This incisive nature of light would attribute therefore the illusion to the perception of phenomena in the sky and space.

However, it is not only the light that produces this illusion, but also the Earth's magnetic field and gravity that make light behave extraordinarily. Double refraction and re-focusing of light was the explanation proposed for the illusion of heaven. Teed's model, however, theorized a stationary sun with which he attributed the notion odd that this was actually a focused point of light emanating from a rotating light double helix at the center of the cosmos. Abdelkader's model receives a orbiting sun rather than a stationary sun. A Italian teacher, about whom there is very little information, Paolo Emilio Amico-Roxas was
apparently became concavist after studying the shape of our Earth. His books are available on the Internet as *The problem of space and the conception of world* and *The supreme harmony of the universe* which present what he called the endospheric field theory.
Around 2010, Steven Christopher aka LSC, publishes a video on YouTube and presents the concave Earth model in a very primary way only with a leaf and a pen. Then in 2014, he began to show 3d images of the system and brings a new element: The octahedron that would be at the center of the Earth would generate terrestrial electromagnetism, which turns out be confirmed by CMOR.

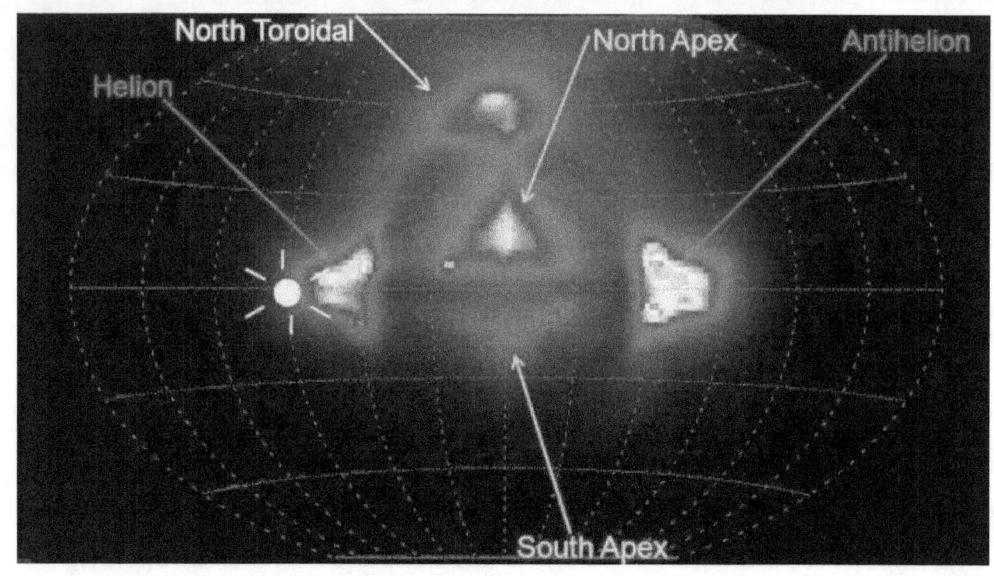

CMOR

The octahedron seen by Steven Christopher says LSC

36

In 2017, I created a website and made personally discoveries (*laterreestconcave.home.blog*) as the real distance stars. I present the reality of our model cosmological and all the physical and mathematics. I also wrote a book under my pseudonym Cyprus Star where all evidence of the reality of our paradigm come together.

Concave Earth means a very clear thing; we let's live in a womb. But which matrix? There is several possibilities but we are sure to be in a globe of perfect dimensions; to the golden ratio.

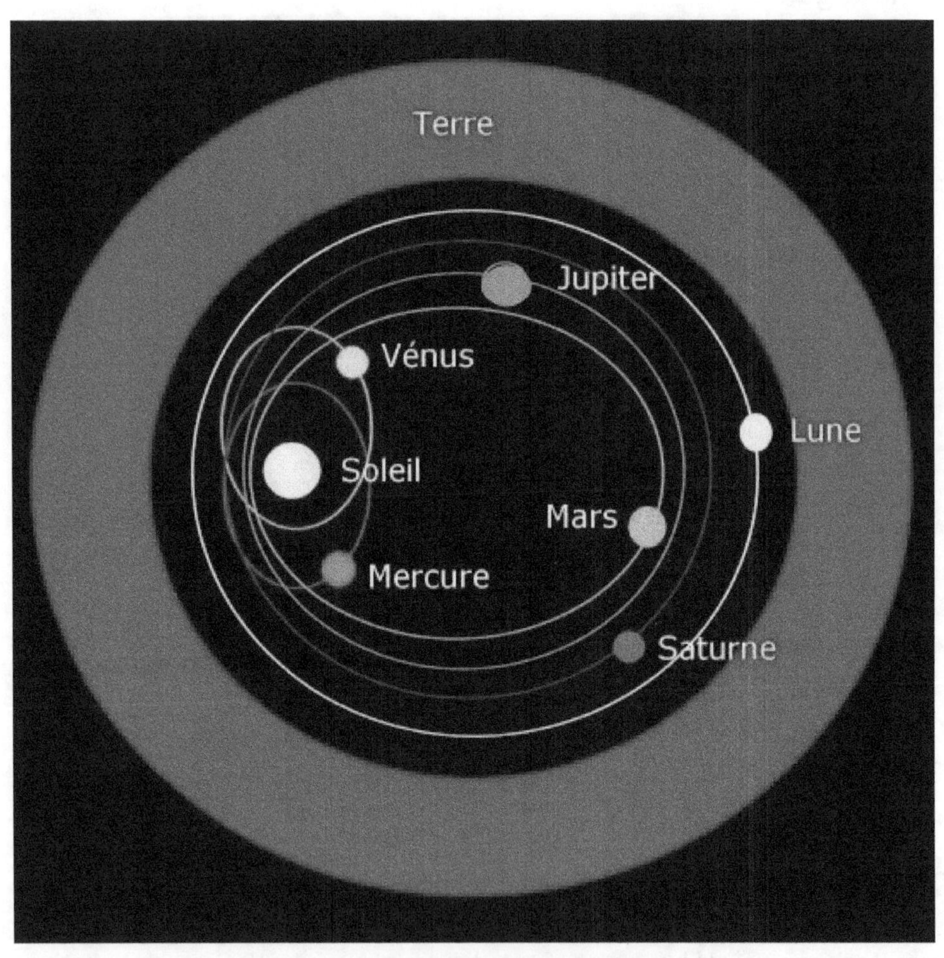

Representation of the real system in the concave Earth. The sphere celestial is at the center of the Earth

6.2 / The Earth has the golden ratio

The golden ratio φ is irrational. He is the only solution

positive of the equation x² = x + 1. It is exactly
(1 + √5) / 2 or approximately 1.6180339887… (more details on
the golden ratio in chapter 7.1 / The golden ratio; the
mathematical proof).

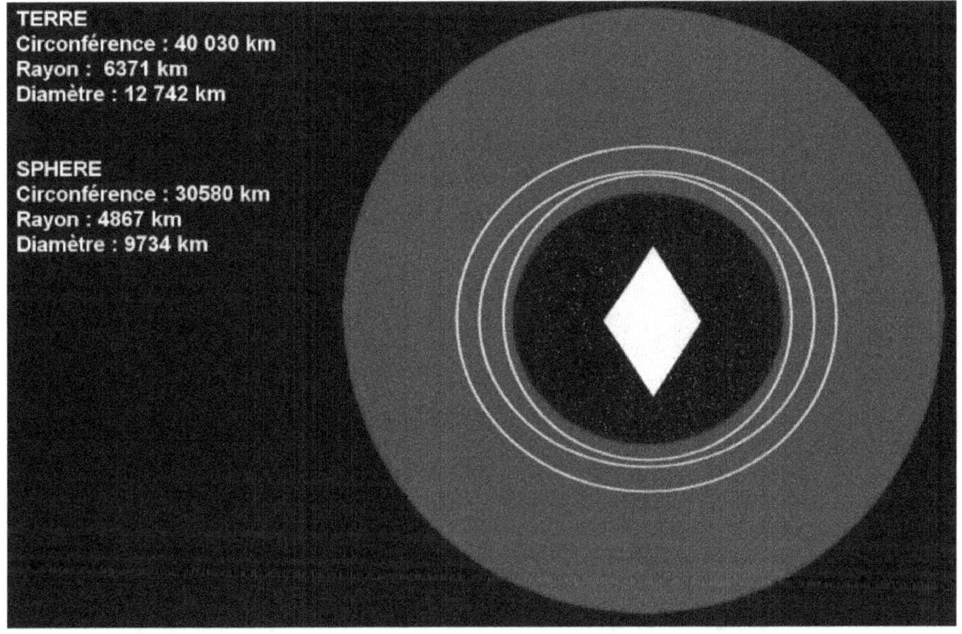

The Earth in blue, the celestial sphere in black with the octahedron at the middle

An irrational number is a number that it is not possible to reduce
as a ratio or as a fraction.
Unlike π, φ is not a transcendent number. φ is a relation naturally
present in many geometric constructions. The Pentagon, for
example, is a safe source to find the golden number. Each
branch of the star is actually a golden triangle. If we divide the
length of the long side by the small we get the golden number
φ. So here we have a report φ in the construction of golden
triangles. But there are 2 triangle levels. And if we compare the
lengths of sides of these triangles from one scale to another is

also φ that stands out! Several peculiarities can be deduced from the equation $x^2 = x + 1$ whose solution and φ and is :

$(1 + \sqrt{5}) / 2$
$φ^2 = φ + 1 ≈ 2.6180339887$
$1 / φ = φ - 1 ≈ 0.6180339887$
$\sqrt{5} = φ + 1 / φ ≈ 1.6180339887 + 0.6180339887 ≈ 2.236067977$

Thanks to the above equations, the golden ratio is certainly the only number for which we can do coincide a geometric progression and a arithmetical progression.
The arithmetic progression is obtained by adding
two successive numbers in the sequence to find the
following.
For example: $0.618 + 1 = 1.618 → 1.618 + 1 = 2.618 ...$
etc.
Regarding our system, the golden ratio is ubiquitous:

Occupancy rate of the octahedron in the sphere:

6,014 km (Octahedron diameter): 9,734 km (celestial sphere diameter) x 100 = 61.78%

Occupation rate of the sphere in the Earth:

9,734 km (celestial sphere diameter): 12,742 km (diameter Earth) x 100 = 76.39%
61.78: 76.39 = 0.8087 x 2 = 1.61
9734 Km (celestial sphere diameter): 1.618 = 6014 Km (Octahedron diameter)
12,742 Km (Earth diameter): 4,867 Km (sphere diameter celestial) = 1.309 x 2 = 2.618
9,734 km (celestial sphere diameter): 6,014 km (diameter Octahedron) = 1.618

This suggests that if our world is perfect, there is no
maybe created not by the chance of an evolution
chaotic but by a creator. It doesn't matter whether it is
a divine entity or a computer program,
someone created this world for a very specific reason.

6.3 / Extraterrestrial research

In a previous book "*UFO Lies, 70 years of disinformation*", I
proved that UFOs were due to sightings of secret military devices
or to other rational explanations. I myself have witnessed a
UFO in 1990 or should I say from an American aircraft.
In terms of the search for extraterrestrial intelligence, a
Australian team published the results of their study
in *Publications of the Astronomical Society of Australia* . These
researchers used the large Australian radio telescope, the MWA,
and analyzed 6 exoplanets and more than 10 million star
systems in the Véla region of our Milky Way. They were looking
for technosignatures, i.e. indicators of advanced extraterrestrial

civilizations. He this is the deepest and broadest investigation in this day. The MWA, therefore, has to look for radio signals, in frequencies similar to FM radio. Result: not one only whisper of alien technology. But is this astonishing? Not really. As for the Fermi paradox who wonders why our civilization does not observe of extraterrestrial civilizations, the answer is simple: because the Earth is closed. If alien there is, this is not in our reality.

7 / Intelligent design

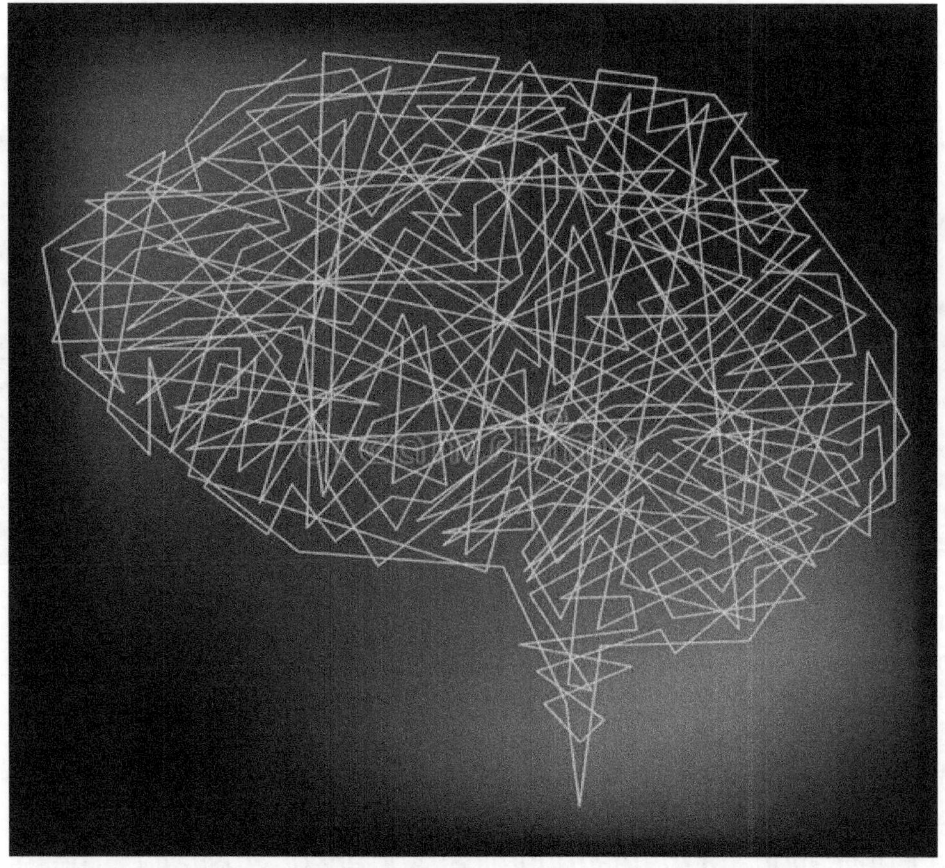

There are two major competing theories concerning
origins of life: evolution and creationism. Evolution represents the opinion of the established science of how the universe has started, and creationism offers the explanation religious. Then

came smart design. The theory of smart design claims that life as we know it we know could not have developed through random natural processes and that only creation intelligence can explain the complexity and diversity that we see today. In 2004, the school board in Dover, Pa. Voted to demand intelligent design education parallel to the evolution in the classes of science. However on December 20, 2005, the judge of United States District Court has ruled that the district school could not follow up because he would violate the constitutional separation of Church and State.

The judge, undoubtedly influenced by the established system, concluded that this was not science and that this teaching could not be dissociated from the field of religious. But in reality there is scientific evidence smart design. The first argument is based on the information scientists find in the cell. The arrangements of the four nucleotides, ACTG, contain specified information and give a meaning for the production and arrangements of protein. The second is a process called correction DNA error (or DNA repair). Our body creates a new DNA constantly in a known process under the name of replication; the DNA double helix is split in half, and if an incorrect nucleotide is deposited, an error correction process detects the anomaly, stops it, removes the wrong nucleotide to put the correct one, then allow the replication process to continue. In addition to repair mechanisms, DNA also contains proofreading processes which ensure the accuracy of the information. All this is produced when messenger RNA transcripts are translated in protein.

The complexity is totally mind-blowing.

7.1 / The golden ratio; mathematical proof

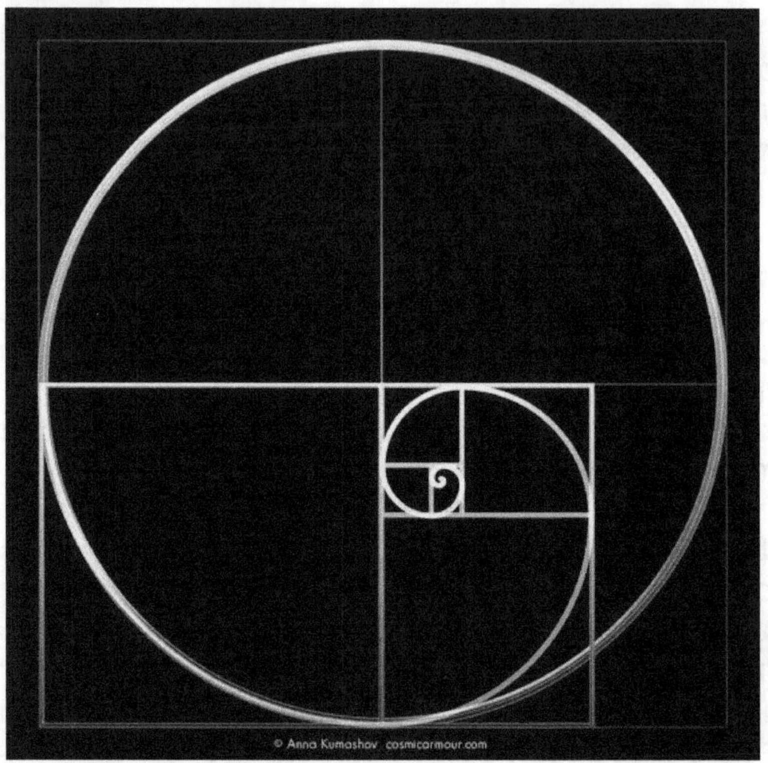

Remember that the golden mean, the golden ratio or the divine proportion refers to the act of cutting a line at a specific point. Point C intersects line AB at divine proportion if the entire line (α + β) is proportional to the longest segment (α) in exactly the same proportion as the most long (α) is at the shortest segment (β).

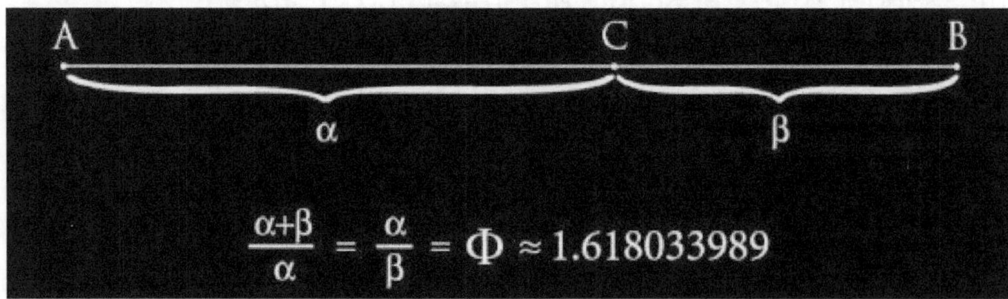

$$\frac{\alpha + \beta}{\alpha} = \frac{\alpha}{\beta} = \Phi \approx 1.618033989$$

The Greek letter Φ (phi) is used to denote the divine proportion. Φ is an irrational number whose the decimal approximation continues indefinitely without repeat. The golden rectangle is a two-dimensional figure where the ratio of the longest side to the shortest side is equal to Φ divided by 1, which reduces to Φ ..

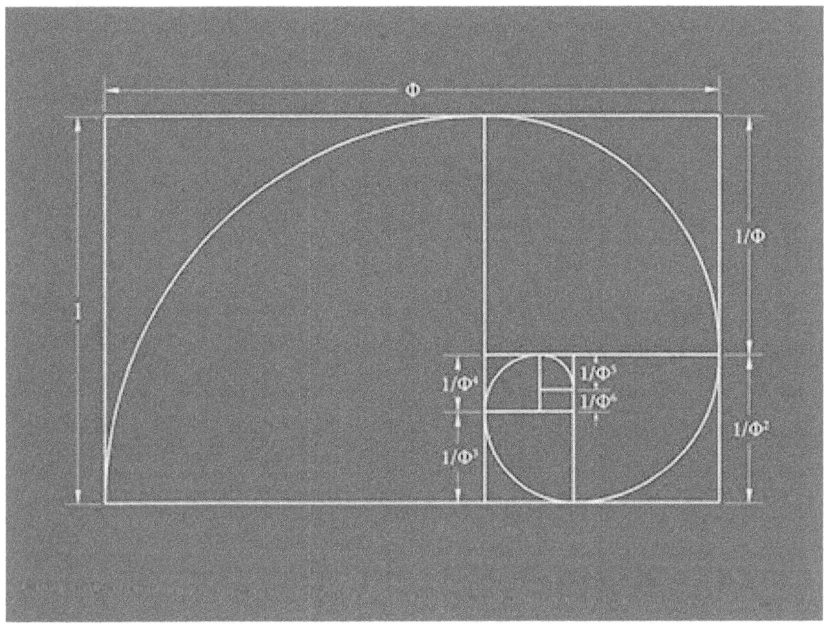

On the right side of the figure, the shorter side is at again subdivided so that what was once the smallest part is now the most large, namely the side with a length of 1. The ratio of the new larger segment (1) to the next smaller one segment is equal to 1 divided by 1 / Φ, which reduces also to Φ. The subdivision continues by reducing always at Φ. The golden rectangle implies a spiral, composed of circular arc segments centered on the corner of each square. The smallest rectangle at the end of the spiral is also a golden rectangle, so that the process of divine subdivision continues infinitely inward. The divine proportion is always the same and does not depend on the length of the line divided.

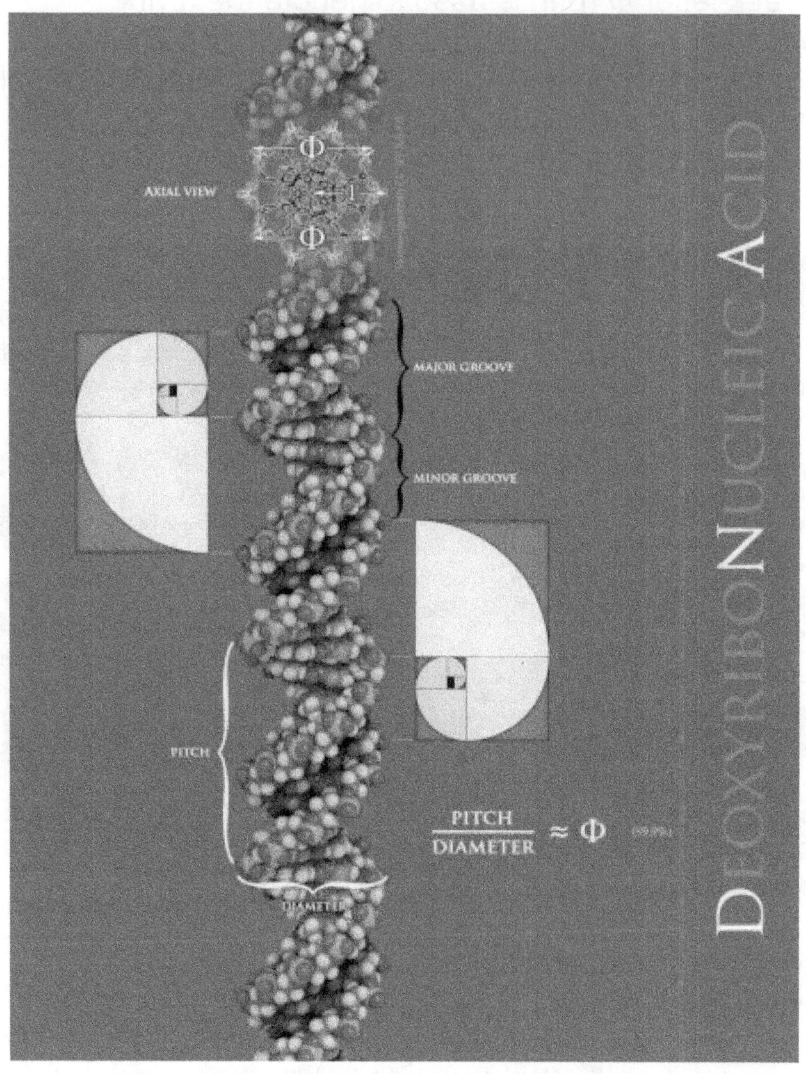

At the molecular level, DNA is divinely proportioned

The top view of the illustration shows that the axis of the structure is arranged as a decagon, with 1 as radius and Φ dimensioned through the molecule. The relationship of the double helix between the major grooves and minor is also divinely proportioned. The overall pitch of the spiral with respect to its diameter is also an expression of Φ (99.9%).
DNA is organized into nucleotide triplets, called

codons. Geneticist Jean-Claude Perez discovered that the frequency of occurrence of codons in the human genome is strongly related to Φ. The code of life is a multidimensional embodiment of the divine proportion.

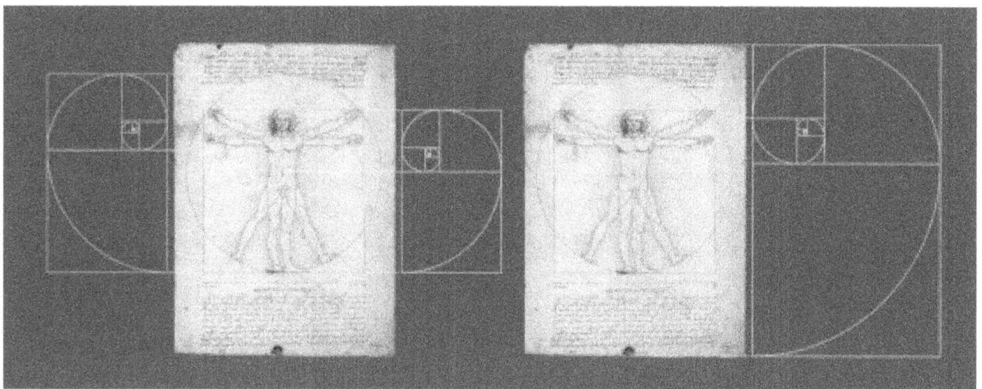

Leonardo da Vinci represented one man at a time in a circle and in a square, based on the old standard of proportion found in the writings of Vitruvius. The divisions primaries in golden rectangles point to the navel and the heart. The navel is the center of the physical body. The heart is the center of the energetic body, with three chakras above and three below. The golden rectangle anchored on the page itself points to the third eye of the idealized man.

The face and the brain

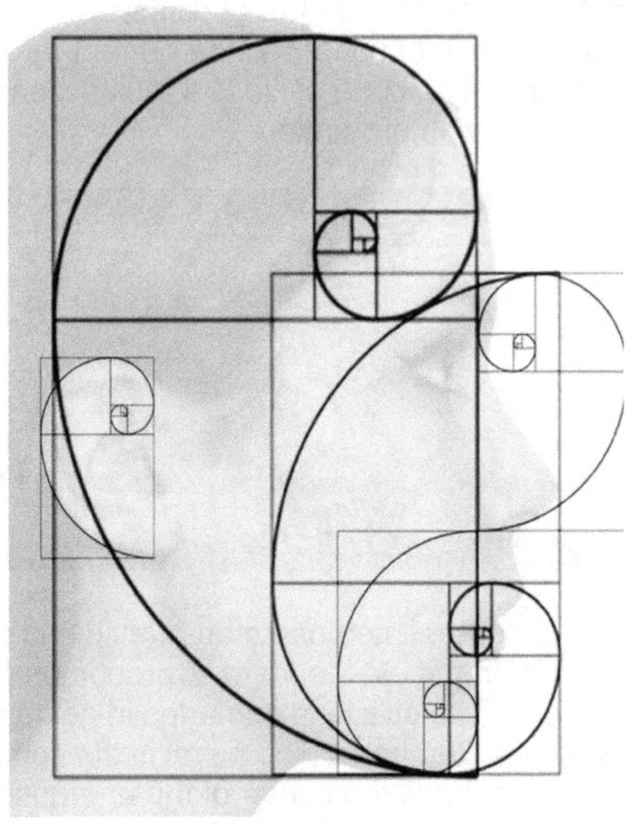

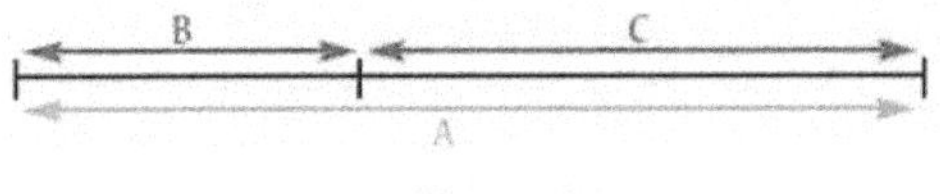

1   Golden Ratio Ø : $\dfrac{C}{B}$ = $\dfrac{A}{C}$ = 1.618...

1: THE GOLDEN RATIO

When a line is divided into two unequal segments, they are in the golden ratio if they satisfy the equation shown (*above*).

2: THE HUMAN SKULL

In the arc drawn over the skull (*right*), the two segments as demarcated by the bregma are in the golden ratio.

Source: Johns Hopkins University

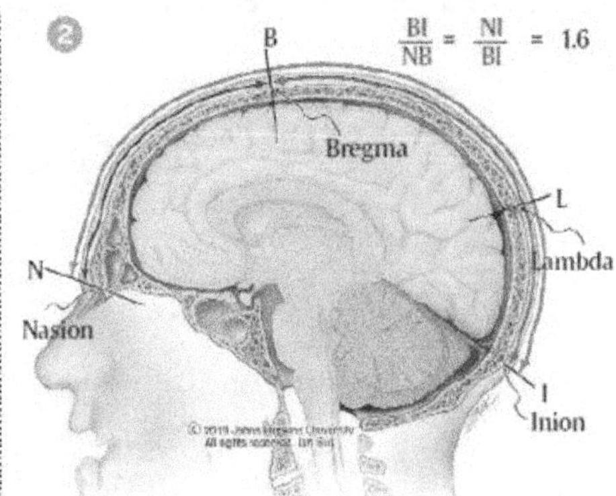

The hand

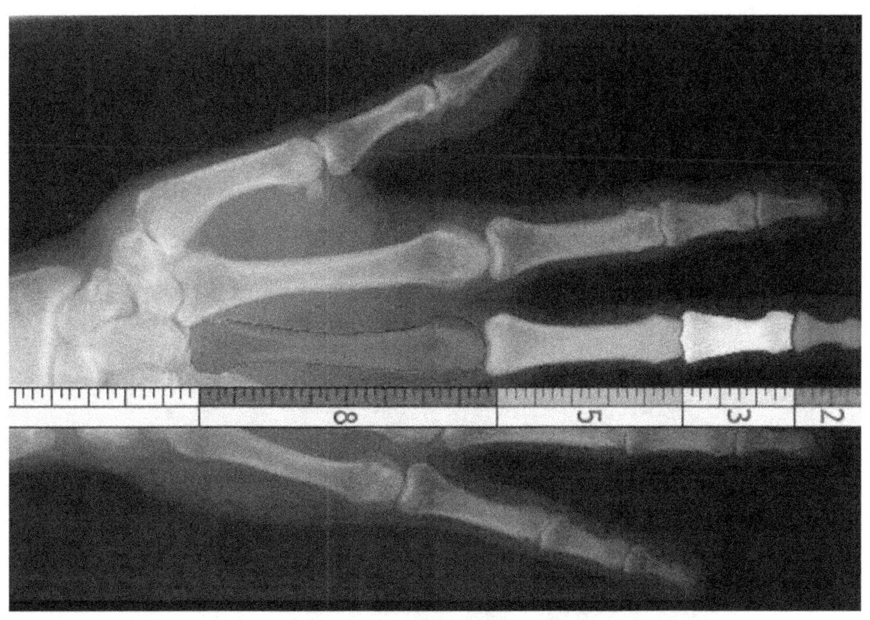

In nature

7.2 / Physical proofs of intelligent design

The skin

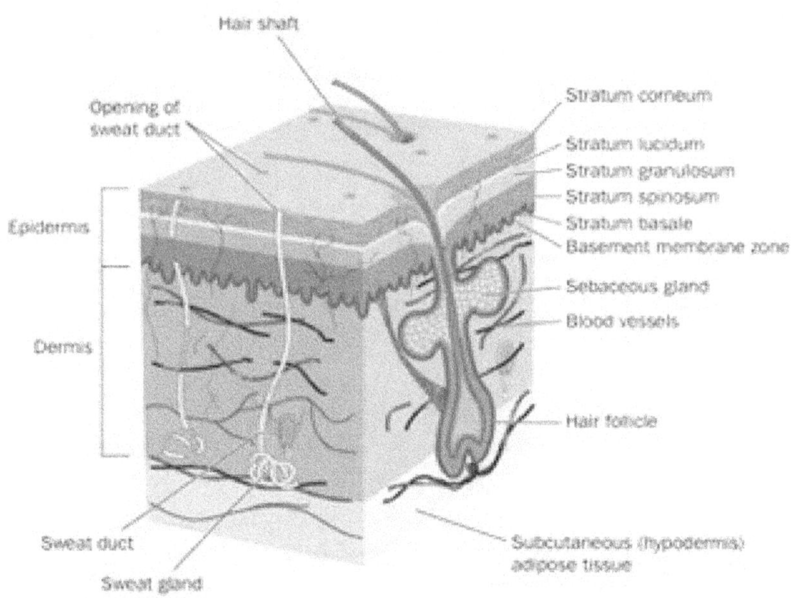

Where can we find 19 million cells, 625 glands sweat, 90 sebaceous glands, 65 hairs, 6 meters blood vessels and 19,000 cells sensory? In a 2.5 cm square of skin human! Human skin is considered to be the largest organ in the body (around 16% of our weight body) and covers an area of 6 square meters. The skin constantly renews itself, excreting more than 30 000 dead cells every minute or 4 kg per year, so that the skin is completely replaced within a year by newly cultured skin cells.

The outer layer of the epidermis is renewed every 28 days. Thus the dead skin cells produce more than half of the dust in our homes! The skin can heal itself by forming scar tissue and if she is exposed to physical pressure, she can forming a callus to provide thickness and toughness additional. Different parts of the body have different types of skin to protect an area particular body. If the skin did not heal of itself, humans and animals would not live for a long time, this healing could not result from evolutionary process over long periods. Without count that the skin contains more than 1000 types of bacteria mainly benefiting the skin by helping it to heal wounds, reducing inflammation and helping the body's immune system fight against infections. The skin removes wastes from the body, including toxins from sweat glands and pores. It regulates body temperature through millions of nerve endings that allow detect sensations such as heat, cold, pain and pressure. The nerves in the skin are connected to muscles that send signals to react quickly to heat or pain. The receptors in the skin respond to pain and to touch. The skin of the fingers and palms is more thick and rougher to allow gripping of objects, otherwise a glass of water would slip out of our hands. Each no one has a unique fingerprint in their skin which is used to identify them. How could this it be a random accident? None of the functions of the skin could not have evolved over millions of years without that humans and animals do not die before they reproduce. All the functions of the skin must have were present during the original creation in order to keep humans and animals alive enough long to be able to reproduce. Therefore, the skin is intelligently designed by a designer.

Blood clotting

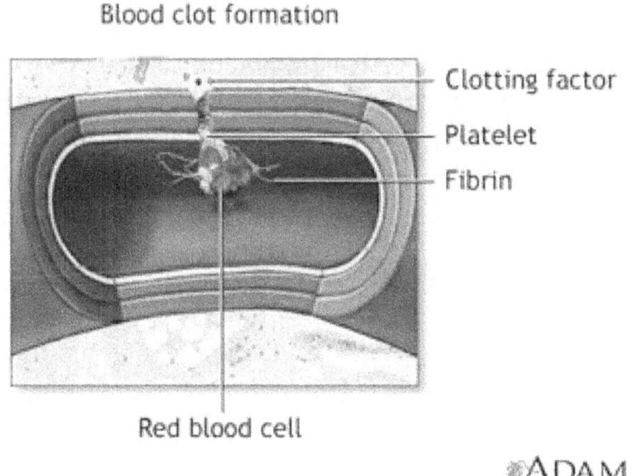

Here is a picture of a cell trapped in a clot. The mesh is formed by a protein called fibrin. When an animal is cut, a protein called Hageman's factor sticks to the surface of cells near the

wound. The bound Hageman factor is then cleaved by a protein called HMK to produce the activation of Hageman factor. Immediately, the Hageman factor activated converts another protein, called prekallikrein, in its active form, kallikrein. Help HMK to speed up the conversion of more factor Hageman in his active form. The Hageman factor activated and the HMK then together transform another protein, called PTA, in its active form. PTA activated in turn, associated with the activated form of another protein called convertin, transforms a protein called the Christmas factor in its active form. The factor of Christmas activated, with the antihemophilic factor changes the Stuart factor in its active form. The latter acting with accelerin, converts prothrombin to thrombin which cuts fibrinogen to give fibrin, which aggregates with other fibrin molecules to form the mesh clot. Blood clotting requires extreme precision. When a circulation system blood pressure is punctured, a clot should occur train quickly or the animal will bleed to death.
On the other hand, if the blood freezes at the wrong time or in the wrong place, the clot can block circulation as it does in heart attacks and stroke. In addition, a clot must stop the bleeding along the entire length of the cut, sealing it completely. However, the coagulation of blood must be confined to the cut or the entire system the animal's blood could solidify and kill it. Through therefore, coagulation requires this system extremely complex so that the clot does not shape only when and only where it's needed.

Cilium (cellular cilia)

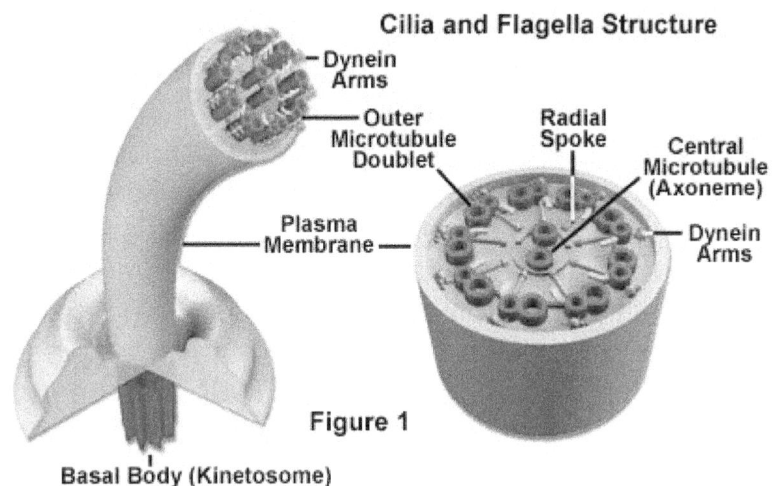

Figure 1

There are irreducible biochemical complex systems. A good example is the eyelash. Eyelashes are hair-like structures on the surfaces of many lower animal and plant cells which can move the fluid on the surface of the row individual cells through a fluid. One eyelash is made up of a bundle of fibers called an axoneme which contains a ring of 9 double microtubules surrounding two central single microtubules. Each doublet outer consists of a ring of 13 filaments (sub-fiber A) fused to an assembly of 10 filaments (sub-fiber B). The filaments of microtubules are made up of two proteins called alpha and beta tubulin. The 11 microtubules forming an axoneme are held together by three types of connectors: the subfibers are connected to the central microtubules by radial radii; the adjacent outer doublets are joined by linkers of a highly elastic protein called nexin; and the central microtubules are connected by a connection bridge. Each sub-fiber A carries two arms, one inner arm and one outer arm, all two containing a protein called dynein.

Experiments have shown that ciliary movement results from chemically powered activation of the arms dynein on a microtubule up a second so that the two slide one over the other. Protein cross-links between microtubules in an eyelash prevent neighboring microtubules from slipping on top of each other for more than a short distance. These crosslinks convert sliding motion induced by dynein in a flexion movement of the entire axoneme. Ciliary movement requires microtubules; otherwise, there would be no strands to slip. He you need a motor, otherwise the microtubules of the cilium would remain stiff and still, and the binders have to pull on the neighboring strands in order to convert the movement of bending slip and prevent the structure from falling disaggregate. Ciliary movement does not exist in the absence of microtubules, connectors and engines. So we can conclude that the eyelash is of a complexity such that it constitutes evidence against Darwinian evolution.

Protein structure

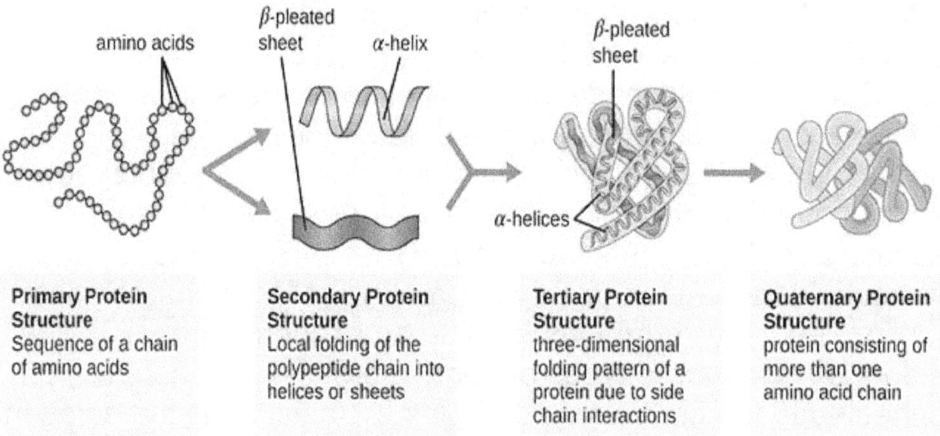

Proteins are large, complex molecules that play a key role in virtually all cell operations. Biochemists have since known long as the three-dimensional structure of a protein dictates its function. Because proteins are such large complex molecules, biochemists categorize protein structure into four levels different: primary, secondary, tertiary and quaternary. The primary structure of a protein is the linear sequence of amino acids that make up each of its polypeptide chains.

Secondary structure refers to arrangements three-dimensional close-range skeleton of the polypeptide chain resulting from interactions between chemical groups that make up its backbone. Three of the most common secondary structures are the random coil, alpha helix (α) and leaf folded beta (β). Tertiary structure describes the form overall of the entire polypeptide chain and the location of each of its atoms in space three-dimensional. The structure and spatial orientation of chemical groups that extend from the backbone of the protein are also part of the tertiary structure. The quaternary structure appears when several chains individual polypeptides interact to form a functional protein complex.

Within the tertiary structure of proteins, the biochemists have discovered compact regions and autonomous which fold up independently. These regions three-dimensional structure of the protein are called domains. Some proteins are made up of a single compact domain, but many proteins have several domains. Indeed, the domains can be considered as the units fundamentals of the tertiary structure of a protein. Each domain has a function unique biochemical. Biochemists refer to the spatial arrangement of domains such as architecture domain of a protein.

Researchers have discovered several thousand distinct protein domains. Many of these domains reproduce in different proteins, the tertiary structure of each protein consisting of a mixed combination of protein domains.

Biochemists have also learned that there is a relationship between the complexity of an organism and the number of unique domains found as a whole proteins and the number of multi-domain proteins encoded by its genome. As much progress as biochemists have made in characterization of the structure of proteins during decades, they still lack a fundamental understanding of the relationship between primary structure (the amino acid sequence) and the tertiary structure and, therefore, the function of protein. In order to develop this vision, they must determine the "rules" that dictate how proteins fold up. Treat proteins as information systems can help determine certain of these rules. Proteins are not just complex molecules, but also carrier systems information. The amino acid sequence that defines the primary structure of a protein is a type of information biochemical with individual amino acids analogous to the letters that make up an alphabet.

To better understand the relationship between the structure primary of a protein and its tertiary structures, the UAB and NIH researchers performed an analysis n-gram out of the 23 million protein domains found in 4800 protein sets species found in all three domains of life.

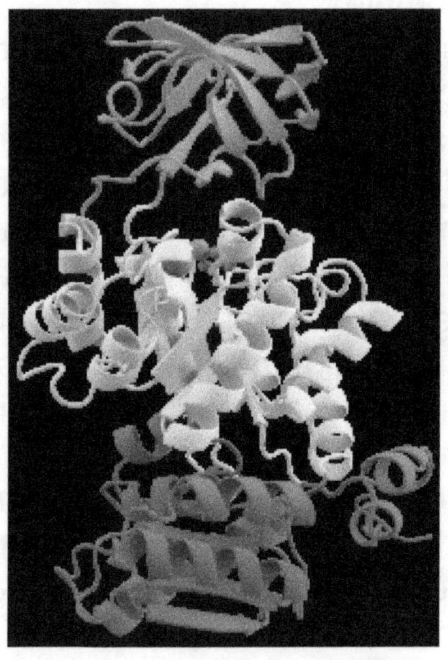

Pyruvate kinase, an example of three domain protein

These researchers point out that an individual amino acid in the primary structure of a protein does not contain information, just like an individual letter in a alphabet has no meaning. In the language human, the most basic unit that conveys meaning is a word. And, in proteins, the most basic unit that conveys biochemical significance is a field. To decipher the grammar used by proteins, the researchers processed adjacent pairs protein domains in the tertiary structure of each protein in the sample. By examining the proteins found in their data set of 4800 species, they discovered that 95% of all possible domain combinations did not exist!

This finding is essential. This indicates that he there are rules that dictate how domains interact. In other words, just like some word combinations never occur in human languages because of the rules of grammar, there seems to be a protein grammar that constrains combinations of domains in protein. The physico-chemical constraints which define the grammar of proteins dictate the tertiary structure preventing 95% of interactions domain-domain imaginable.

In thermodynamics, entropy is often used as a measure of the disorder of a system. The information theorists borrow the concept entropy and use it to measure the content informational of a system. For theorists of information, the entropy of a system is indirectly proportional to the amount of information contained in a sequence of symbols. As the information content increases, the entropy of the sequence decreases, and vice versa. Using this concept, UAB and NIH researchers calculated entropy combinations of protein domains. In the human language, entropy increases as the vocabulary increases. This makes sense because, at as the number of words Increases in a language, the probability that combinations of words random contain decreasing meaning. Of the Similarly, the research team found that the entropy of the protein grammar increases at as the number of domains increases The human languages all convey the same quantity information. That is, they all display the same entropy content. Information theorists interpret this observation as an indication that a universal grammar underlies all human languages. It is fascinating that researchers discovered that the protein languages of prokaryotes and eukaryotes all display the same entropy level and, therefore, the same content of information. This relationship holds despite the diversity and differences in the complexity of the organism in their dataset. By analogy, this discovery indicates that a universal grammar exists for protein. Or to put it another way, the same set of physicochemical constraints dictate how protein domains interact for all organizations.

This study also illustrates how it can be successful in treating biochemical systems as information systems. The researchers conclude that the similarities between natural languages and genomes are apparent when domains are treated as functional analogues of words in natural languages. It is this relation which indicates the role of a creator in the origin and conception intelligent life. The arguments in favor of creation are reinforced when we recognize that it is not simply the presence of information in biomolecules that contributes to this version of a revitalized watchmaker analogy. Vigor additional comes from the discovery of the researchers of UAB and NIH that the mathematical structure human languages and biochemical languages is identical. The discovery of a protein grammar means that there are physicochemical constraints on the structure of proteins. It is remarkable to think that the tertiary structures of proteins can be fundamentally dictated by the laws of nature, instead of being the result of an evolutionary history historically contingent. The discovery of a protein grammar reveals that the structure of biological systems may reflect deep principles and underlying which derive from the very nature of the universe itself. These structures are precisely those that life needs to exist. This can be the proof that our universe was designed from scratch.

The human eye

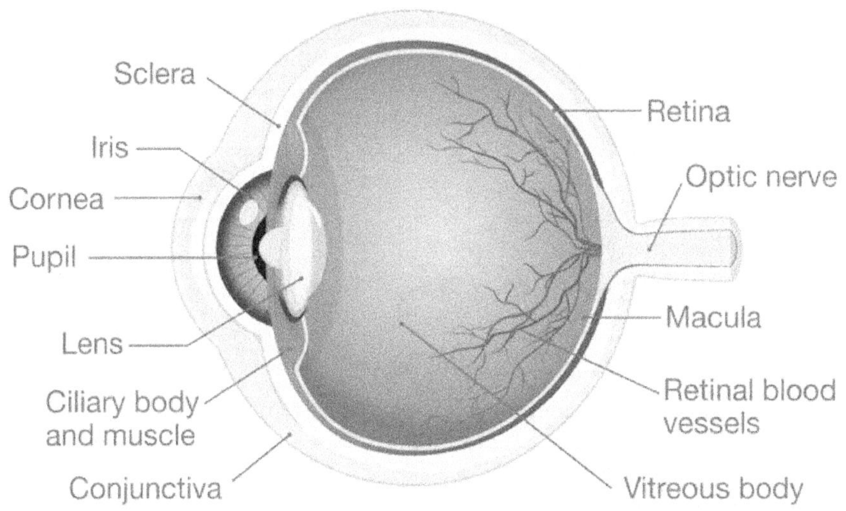

According to National Geographic, the eye was formed by mutations according to the theory of natural selection. The concern with natural selection is that it does not no mutation or change that is not immediately functional, therefore assume that the eye and all its complex parts come from millions of mutations, is impossible. A mutated retina would not survive not without the lens, and a mutated lens would not survive without the retina. Paul Ehrlich pleads for a design intelligent:
" Because mutations are random compared to needs and because organizations integrate generally well in their environment, mutations are normally either neutral or harmful; they are only very rarely useful, all as a random change made by pressing a screwdriver in the guts of your computer will rarely improve its performance."

Smart design says parts of the universe and nature are best explained by a cause smart. The eye fits this definition exact. Even Charles Darwin said: "*To assume that the eye, with all its inimitable artifices to adjust the focus at different distances, to admit different amounts of light and to correct spherical and*

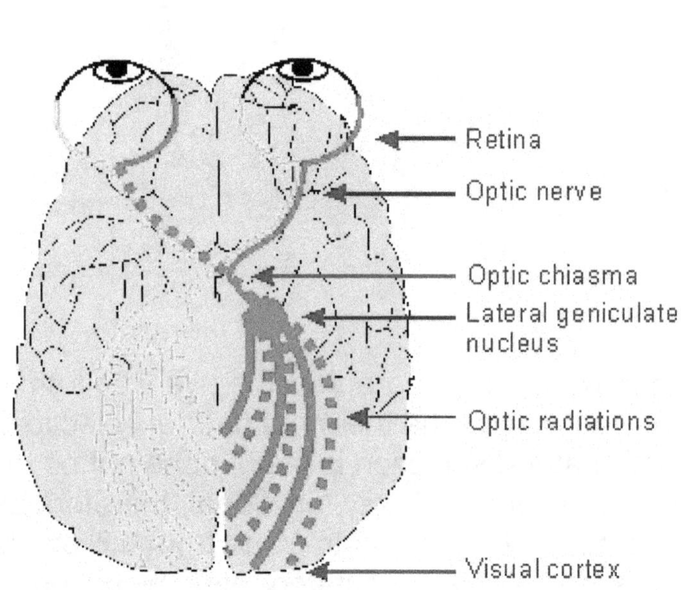

Finally, the eye shows an intelligent design because its immense structures and systems could not have been created by a series of random events. The eye was planned, and was done by a creator. He even thought of tears to wet the eyes and rid them of any dirt inside.

Other examples

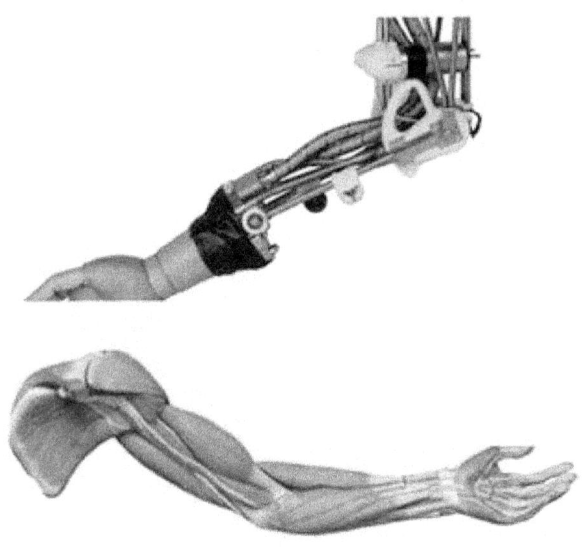

When we compare the human arm to that of a robot, striking similarities appear.

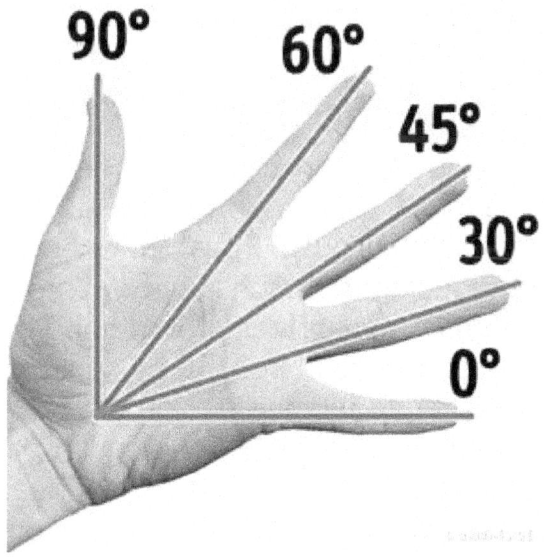

The hand is perfectly designed; a right angle is formed when it is fully open.

Fossils

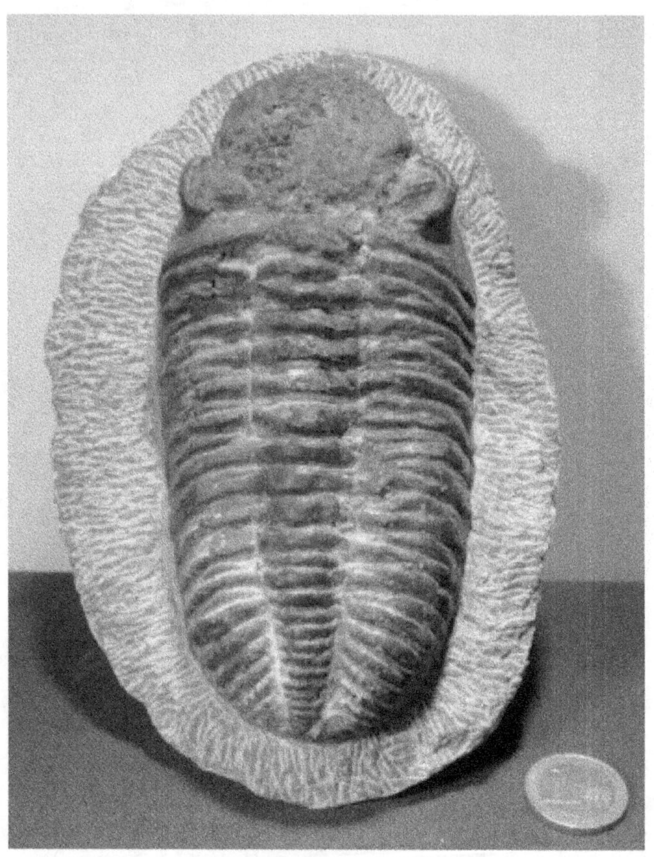

Fossil trilobites exhibit all the attributes of a
creature designed for a purpose. Fossils present
features that seem designed, much like living organisms. The
real cause of the fossils can only be determined by testing
agents known causal agents to see if they can produce the
specified complexity evident in the design of fossils. Unintelligent
causes such as random variation of mutations or the
deterministic law of natural selection have not been shown to be
capable of produce a specified complexity, let alone a new and
more specified complexity. More precisely, natural selection
(which can only select what is already created) has no creative
power. The only ones things that can be "selected" are already

existing things. Therefore, in the Darwinism, any "creation" of
new information DNA encoding new body designs
as evidenced by the fossils must be produced by non-intelligent
deterministic physics. Each random mutation of the genome of
an existing body plan must be directionally beneficial for produce
a new beneficial body plan. But, as discussed above, no natural
process unintelligent has only been shown to produce a new
beneficial morphology, or new content beneficial information in
the genome, or new cash. Therefore, the process capacity
Darwinians to produce the obvious specified complexity in fossils
is seriously questioned.

8 / The shape of the universe

The study of the curvature of the universe is an area of
constantly evolving research in cosmology. These
recent years, the data collected by observation missions such as
WMAP and Planck have showed a locally zero curvature of the
Universe, indicating that the latter is therefore certainly flat. This
model of the Flat Universe is now integrated into the standard
cosmological model, but an anomaly from data collected by the
space observatory Planck in 2018, concerning the cosmological
diffuse background, could be interpreted as the sign of a
universe spherical closed. According to an international team
astronomers from the University of Manchester to United
Kingdom, conclusions force us to rethink radically the current
model. The key in the determination of the curvature of the
Universe lies in the how gravity bends the path of light, an effect
predicted by Einstein and called the gravitational lens.
Data from the Planck satellite, those for 2018 in particular, show
that the CMB undergoes a lens effect gravitational more
pronounced than expected. The Collaboration Planck called this
anomaly Alens, and this has not yet been resolved, but the team

believes that one explanation could be the curvature of the universe.
The study was published in the journal *Nature Astronomy*.

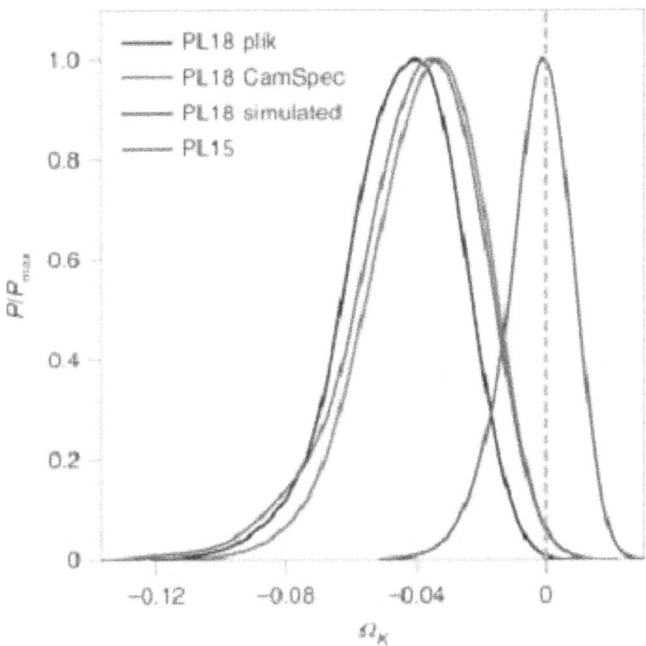

The researchers showed that the anomaly in the cosmic diffuse background spectrum could be interpreted as a sign of a closed universe (blue). Credits: Eleonora Di Valentino. 2019

" A closed universe can provide a physical explanation for this effect, the spectrum of the Planck CMB pointing now towards a positive curvature greater than 99% of confidence. Here we further investigate the evidence of a closed universe collected by Planck, showing that the positive curvature naturally explains the amplitude abnormal lens effect, " the researchers write. Astrophysicists Steven Gratton and George Efstathiou University of Cambridge also analyzed the Planck data from 2018 and highlighted signs of curvature.

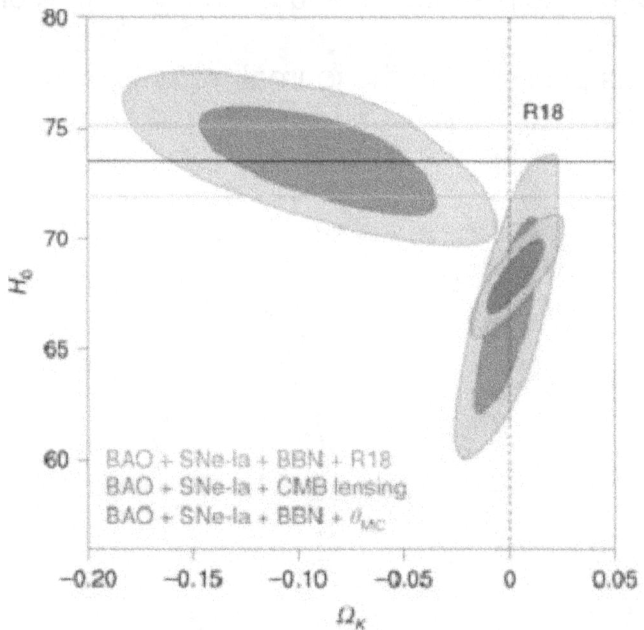

The interpretation of the authors of the article (red and blue) is incompatible with current data from the various cosmological observation missions (gray). Credit: Eleonora Di Valentino. 2019

The French philosopher René Descartes once said:
*"It is possible that I am dreaming right now and that all
my perceptions are wrong. "* The nature of reality has
been meditated on by philosophers for millennia. Although Plato
or Descartes did not think probably not at the matrix, many
people have been able to note that simulation theory is an
allegory of the modern cave or the hypothesis of the evil
demon. The theory of a holographic universe was first introduced
in 1997 by the physicist Juan Maldacena. The latter postulated
that the gravity comes from thin vibrating strings moving in nine

dimensions of space and one dimension of time, 'real' life existing in a gravity-free universe. He be aware that when quantum gravity is combined in quantum mechanics, symmetry is impossible.

The hologram is a three-dimensional image encoded on a two-dimensional surface. The universe is built of the same way, the higher dimensional part coded on a part of lower dimension more flat. So only one part is tangible; the surface lower dimensional on which the hologram is coded.

9.1 / Anti-de Sitter space-time

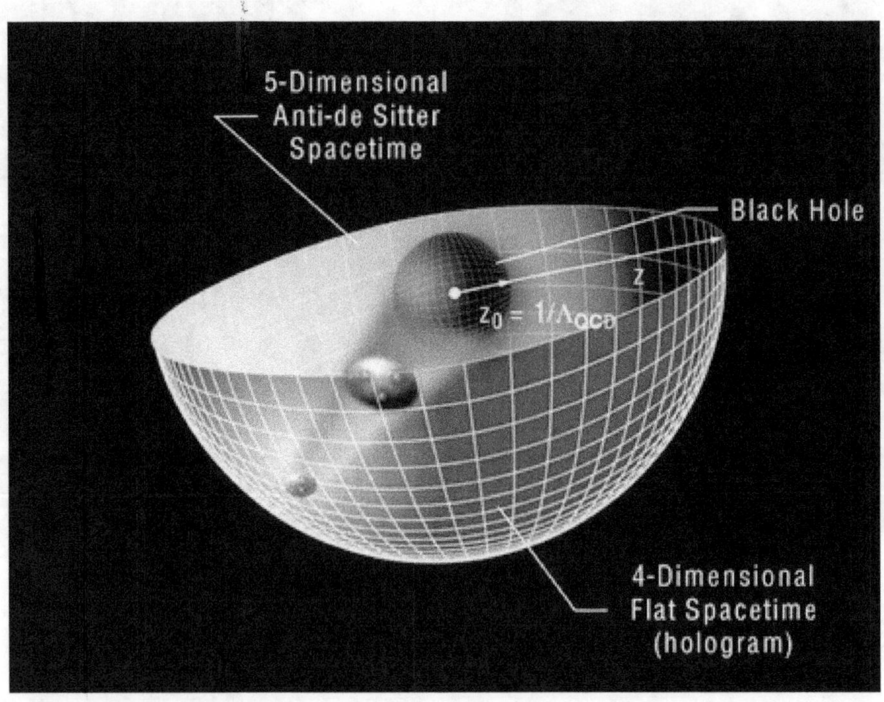

An anti-sitter space-time 3D representation of a holographic universe, characterized by a space-time with negative curvature whose edge is a space-flat time in which quantum field theory applies conforms while inside the string theory applies including gravity. It even forms safe black holes resulting from the interaction of particles on the edge. © Alfred Kamajian

In the theory of the holographic universe, the physical laws are different within a volume and its surface, but equivalent by projection holographic. This is how our universe could be a hologram where the physics of the edge would be physics quantum, and the physics of the interior would reveal, in a kind of illusion, the force of gravity.

To be more precise, the physicists of the theory holographic considers a space-time whose curvature is negative. The border of this space-time anti-de Sitter is a flat (zero-curvature) space-time.

On this frontier, we use the quantum theory of particles and gravity does not exist. The latter, in a holographic universe, is entirely equivalent to a quantum theory of gravity and matter, like string theory. The interaction of particles on the edge produces phenomena inside: strings, gravity, even ephemeral black holes emitting a warm radiation.

In other words, gravity emerges naturally, in a universe with four dimensions of space, particle physics in three dimensions of space (which is our quantum physics). What we call the gravity would therefore be the consequence of the geometry of our universe.

Exactly as in our concave Earth where space is at its center.

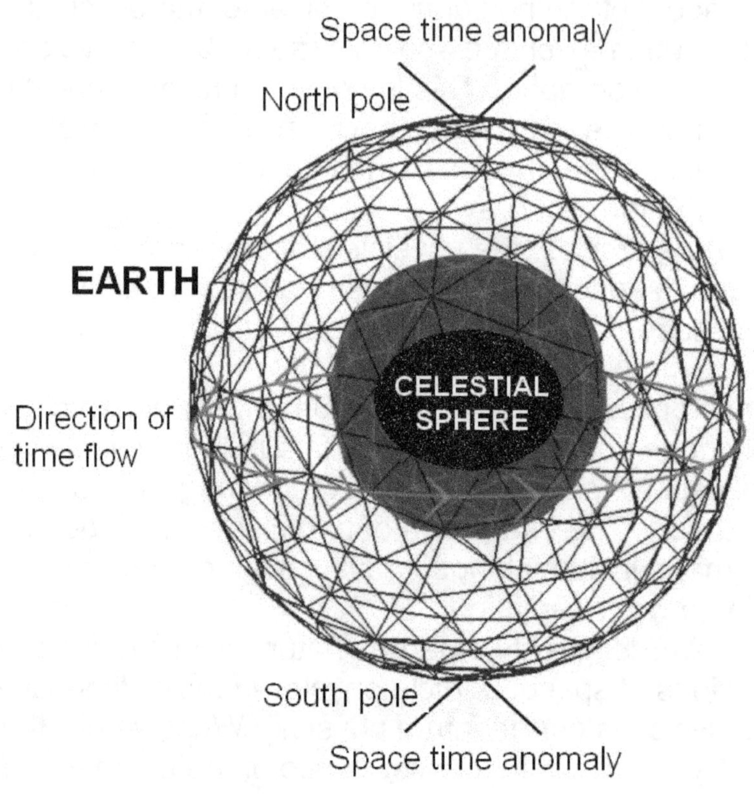

Space time in Earth – 4 dimensions

On the diagram I wrote that there were anomalies in the respective poles. Indeed, it has been proven that the Earth was open, an anomaly having even been found in Antarctica in 2019. The news was picked up by the newspapers which for the most part have headlined: " *The proof of a parallel universe* ", while beyond a parallel universe, this discovery proves the opening of the Earth to the poles.

9.2 / Parallel universes

Cosmology. Evidence of a parallel universe discovered in Antarctica?

SCIENCE & TECHNO ⟩ ALARM CLOCK ⟩ **NEW SCIENTIST - LONDON**

⟨ Published on 05/03/2020 - 06:08

Another related topic of quantum physics is the theory of parallel universes. There is a graph directed by several universes which branch out each time that we make a decision, which results in different deadlines. In which of these universes are we we plugged in? it may have to do with who is most optimal - which means that these universes can exist or not as real physical realities.

The Minimax algorithm examines the possible futures, calculating which one is the most optimal for a video game. Physicist Fred Alan Wolf explains that information about these possible futures reaches us in the present and that we send a wave of offer in the future, which interacts with the waves of supply coming from the future to the present. The possible future towards which where we go depends on the choices we make and how these two waves overlap or cancel each other out. Probable futures send information in the present tense and we choose consciously the path to follow! The physicist Thomas Campbell, in his 2003 book, *My Big TOE* (Theory of Everything) also suggests that there is a function fundamental and that we are essential in a computer universe which forks the possibilities and uses an evaluation function, just like a video game!

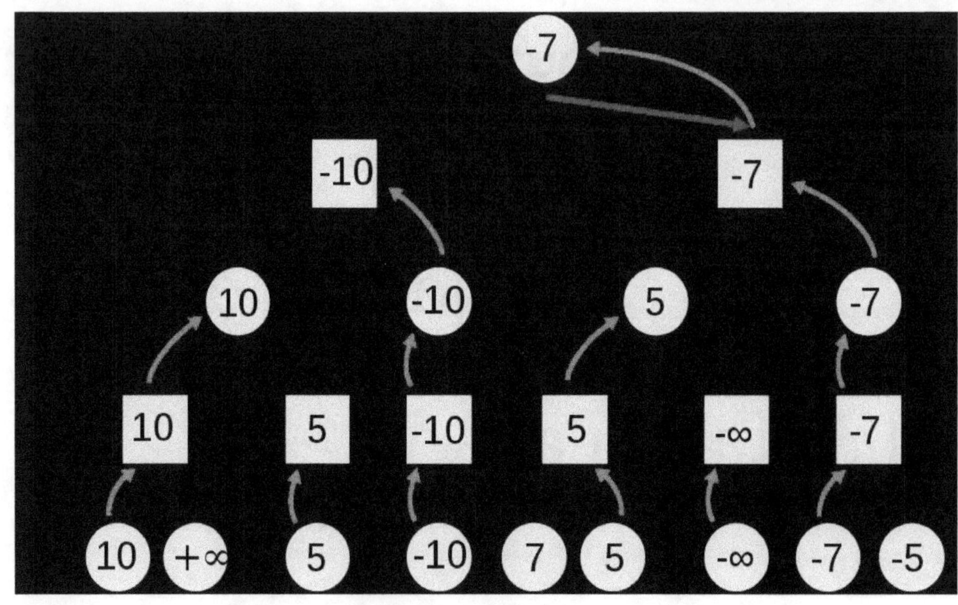

ANITA

In January 2020, ANITA (transitional branch Antarctic impulse) flew and detected many cosmic rays. Two events indicate that they came from the ground. Indeed, the signals detected came from particles resembling ultra-high energy neutrinos moving from the bottom up. But for scientists, the rays cosmics should not do this in large numbers, leaving the source of these signals a mystery. For the scientific establishment, one possibility is that ANITA detected radio waves emitted by a particle not taken into account in the standard model. The team claims that new detections of these strange signals are necessary before drawing definitive conclusions about their origin.

Ignorance of the true cosmological model

For the whole world it is accepted that we live on a globe and the sun is 150 million kilometers. But I have proven since 2017 through numerous articles and videos that the Earth was indeed a globe but that we do not live on it but inside. The sun is actually very close, at around 4000 km. This is called the concave Earth theory (and not the Hollow Earth which is a model wrong). In our real system, space is actually a celestial sphere located at the heart of the planet.

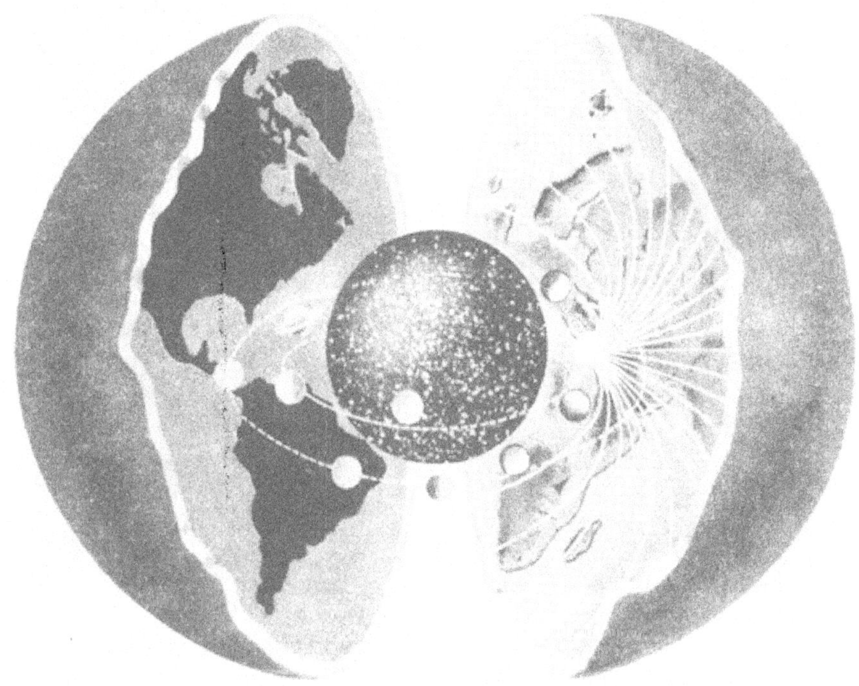

So our little cosmos is suspended in the air thanks among other things to the strength of the charge field positive. In reality the Earth is open to the north poles and South. This is the reason why there are aurora borealis in the northern hemisphere and aurora australes in the southern hemisphere. It is therefore not amazing that scientists found particles springing from the ground of Antarctica during ANITA flights.
Maybe these neutrinos are even coming from outside
of the Earth.

The model of a convex Earth is wrong. It's science fiction backed by NASA that conceals reality. Their image is a CGI most of the time

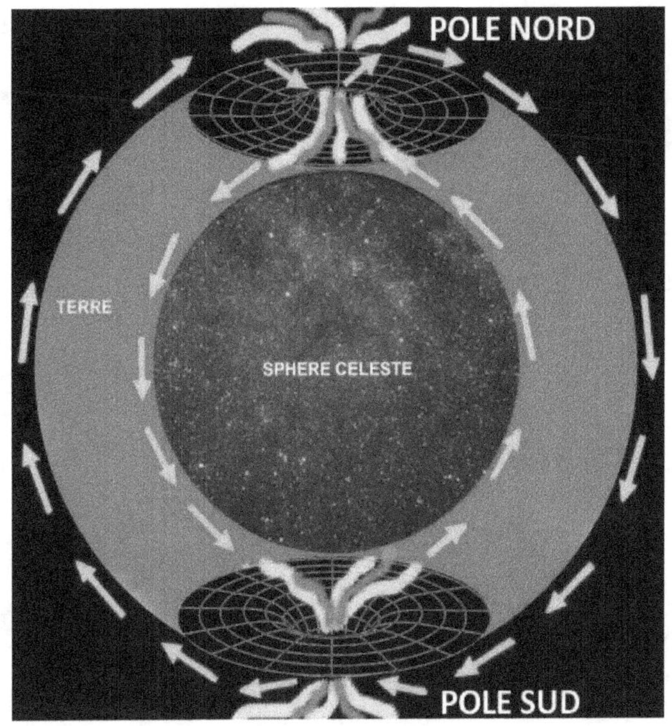

The proton charge continuously crosses the poles to circulate inside the Earth.

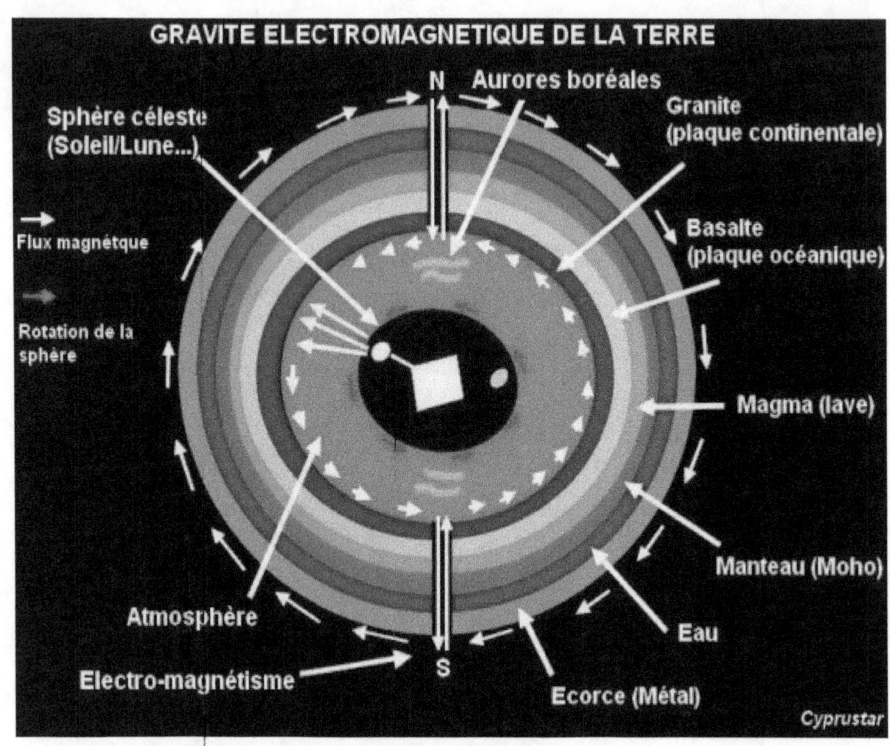

Earth layers

9.3 / Objective reality does not exist

Our brain mathematically constructs reality objective by interpreting frequencies, it is a hologram wrapped in a holographic universe.

These are essentially the theories of Bohm and Pribram that allowed us to look at the world in a way new.

The latter were the main supporters of this great theory; they worked independently and in completely different directions, but they are reached the same conclusions. The two scientists were unhappy with standard theories that do not could not explain the various phenomena encountered in quantum physics and the puzzles related to neurophysiology of the brain. An experiment remarkable project carried out in 1982, by a team of research directed by the physicist Alain Aspect at the Institute of Theoretical and Applied Optics, in Paris, demonstrated that the network of subatomic particles that make up our physical universe - the fabric of reality itself - has what appears to be a undeniable holographic property. Aspect and his colleagues Jean Dalibard and Gérard Roger discovered that in some conditions, subatomic particles such as electrons are able to communicate

instantly between them regardless of the distance between them. Little does not matter if they are 10 feet or 10 billion apart of kilometers. One way or another, each particle always seems to know how to join the other. The problem with this feat is that it violates what was thought Einstein on the fact that no communication can travel faster than the speed of light. Being given that travel faster than the speed of light is tantamount to breaking the barrier of time, this perspective intimidating has prompted some physicists to try to find elaborate ways to explain the Aspect discoveries. Finally, he inspired others
to offer even more radical explanations, all based on the assumption that objective reality does not exist not, that despite its apparent solidity, the universe is a Huge and wonderfully detailed hologram.
Karl Pribram realized that the objective world does not exist the way we know it or we see it.
He claimed that our brain is able to build objects and David Bohm even concluded that we let's build space and time.
There is a fundamental principle in this mysterious science, it is called the quantum superposition which says that, just like waves, two quantum states can be added together and the result will be another valid quantum state. You know the cat of Schrödinger? The hypothetical cat who is both alive and dead, so a cat in a state of quantum superposition. However, as soon as an observer checks animal, superposition state magic disappears and an observer sees a dead cat or living. Many real experiments have shown the principle of superposition. The best known is a double slit experience.
To prove its nature as a wave, a scientist drew a pure wavelength light through a leaf with two slits, as the light passed through the slits, it was divided into two distinct waves.
About a century later, scientists made a quite similar experience, but this time, the experiment involved an electron beam gun which shot electrons through the double-slit device.
If the particles were sent one by one, it would be resulted in the appearance of a single particle on the screen, as

expected. Remarkably, an interference pattern appeared when these particles were able to accumulate a by one. The experiment, which was then carried out on whole atoms and even molecules, has shown that a particle can behave like a wave, the phenomenon has been called wave-particle duality. As soon as we take a look at the particles, they stop behaving like waves and become particles.

In 1961, the physicist Eugène Wigner, winner of the Nobel Prize in Physics, described a thought experiment highlighting one of the paradoxes of mechanics quantum. Experience shows how nature strange universe allows two observers, say Wigner and his friend, to experience different realities.

Recently, physicists have found that recent advances in quantum technologies had allowed to reproduce the Wigner test within the framework from real experience.

Massimiliano Proietti from Heriot-Watt University of Edinburgh and some colleagues claim to have performed this experiment for the first time: they have created and compared different realities. Their conclusion is that Wigner was right: these realities can be made irreconcilable, so that it is impossible to agree on objective facts about an experience. The results have been published on the server arXiv prepublication while waiting for the process of *peer review*.

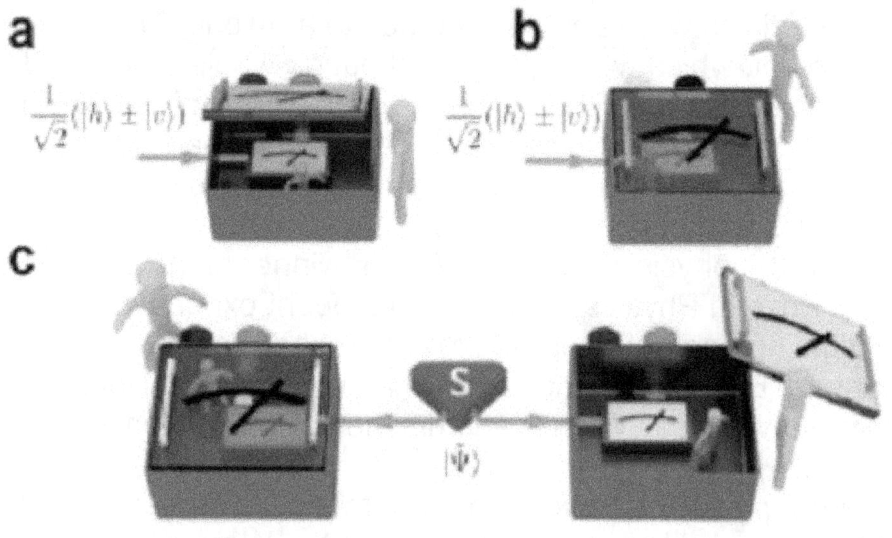

Diagram of the experiment of Wigner and his friend. a) Wigner's friend performs a measurement on the quantum system in a state of layering inside the box. b) Outside the box, Wigner describes his friend and the associated quantum system as entangled. c) Extended version of the experiment, where an entangled state is sent to two different laboratories, each involving an experience and a friend

Wigner can even perform an experiment for determine if this overlay exists or not. It's a kind of interference experiment showing that the photon and the measure are well superimposed. From the standpoint de Wigner, it is a fact, the superposition exists. And this fact suggests that a measurement may not have taken place. But this contrasts with the point of view of the friend, who actually measured the polarization of the photon and has it recorded. The friend can even call Wigner and say that the measurement has been carried out (provided that the result is not is not revealed). So the two realities are in disagreement. " *This calls into question the status of goal facts established by the two observers,* " says Proietti. Caslav Brukner, from the University of Vienna in Austria, has imagined a way to recreate Wigner's experience by laboratory using techniques involving the simultaneous entanglement of many particles. The feat that Proietti and his colleagues have made is to achieve this. " *In a state-of-the-art experience technology involving 6 photons, we realize the scenario imagined by Wigner* " explain the researchers.

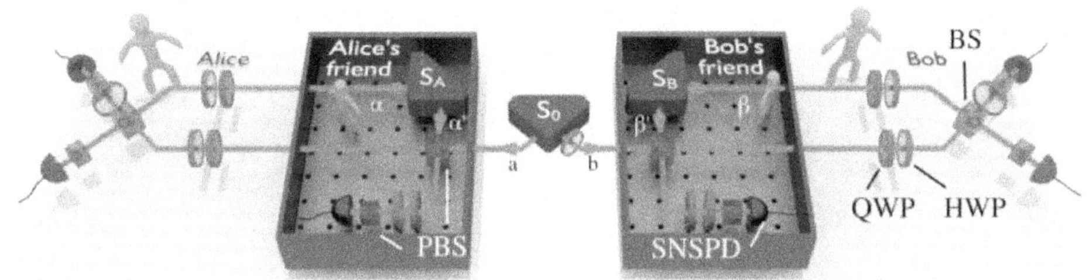

Diagram of the experience Credits: Massimiliano Proietti

They use these six entangled photons to create two alternate realities: one representing Wigner and the other representing Wigner's friend. Wigner's friend measures the polarization of a photon and records the result. Wigner then performs an interference measurement for determine if the measurement and the photon are in overlay. The experiment produces a result without ambiguity. It turns out that the two realities can coexist, even if they produce irreconcilable results, just as Wigner predicted.

This suggests that objective reality does not exist.

9.4 / The pixelated universe

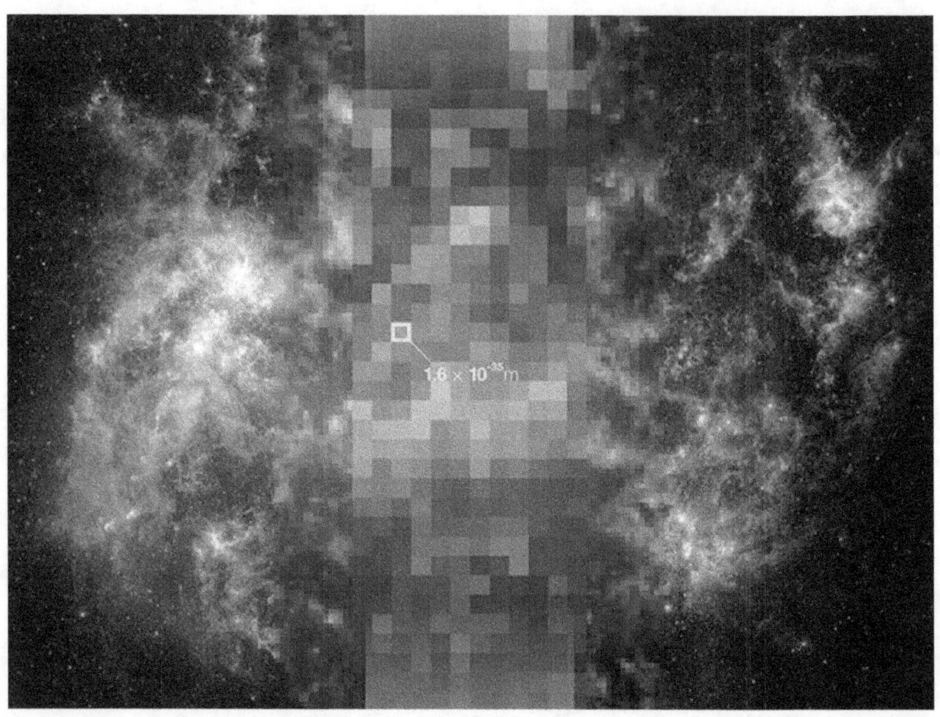

The most basic aspect of the universe does not seem
be matter or energy, but information. Fundamental elements that
make up the world and that physicists call quarks cannot be
divided into smaller parts and combinations of these 12
fundamental particles literally make up every piece of matter we
can observe. Quarks are available in 6 flavors: high, low, high,
low, strange and charm and each of the 6 quarks also has an
opposing or anti-quark particle which brings the total number of
possible quarks to 12. These 12 indivisible pieces of matter
make up the universe. It's oddly similar to pixels in a game video
or an image on the screen which, when viewed up close, does
not are just tiny colored boxes that make up a virtual
world. Energy carriers like photons, are also described as
quanta, this which means that the energy is transmitted in small
pockets of energy instead of a constant flow.
Our universe has been shown to be pixelated, and as it had a
start it is also considered as finished. This would then mean that
our universe is program. Quarks and leptons, constituting the

material, are incredibly small. Even the biggest quarks only measure about an attometer (a billionth of a billionth of a meter) in diameter. But zoom in closer - a billion times closer – beyond zeptometers and yoctometers, where the units are short of names. Then go on, a hundred million times smaller still, and you've finally hit rock bottom: this is Planck's length, about $1.6 \times 10^{3-5}$ meters, considered by physicists to be the longest short possible in the universe. Beyond this point, they say, the very notion of distance no longer makes sense.
One of the great discoveries of the 20th century was that small scale, many physical properties, like angular momentum and energy, cannot take some discrete values, or quanta.

Pixelated Dark Energy

Five scientists studied the phenomenology of a string construction with dark energy mechanically stable quantum. The construction is holographic in the sense that 4D space-time is generated from pixels from five branes enveloped over five metastable cycles of compactification. The cosmological constant as
$\Lambda \sim 1 / N$ in the number of pixels. The sudden onset radiation triggers an increase exponential of the number of pixels. Dark energy has a time varying equation of state with $w = -3\Omega m, 0 (1 + w0) / 2$, which is compatible with current limits, and could be further limited by future data versions. The pixelated nature of the universe also involves a big cut on the angular power spectrum of observables cosmological.

9.5 / Bugs in the matrix

Silas Beane, nuclear physicist at the University of Washington to Seattle, move that we can be able to find previously overlooked bugs by discovering the mathematical structure used to build our simulated reality. He explains that the scientists in their field use a set of lattice-shaped coordinates to simulate the behavior of subatomic particles. If our reality is built on top of a network, there would be a fundamental rudeness, because there could be no in our simulated universe no detail smaller than the simulation resolution. Even though the resolution limit is too small for us to observe directly, we can detect it experimentally. S. Beane proposes that a network of simulation could affect the behavior of particles ultra-energetic called cosmic rays, affecting their orientation and maximum intensity.

Instruments like the Telescope Array, a huge network of 500 detectors scattered in the desert of Utah, watch cosmic rays. Detectors have already discovered particles up to 100 quintillion times more energetic than visible light. The team de Beane calculated that the distribution of most energetic cosmics rays should present a break in symmetry. Something similar to such a limit appears in the data experiments on the most energies confirming a simulated universe with limited computing

resources. Beane keep believing that there are still other
possibilities for simulated discover the simulators!

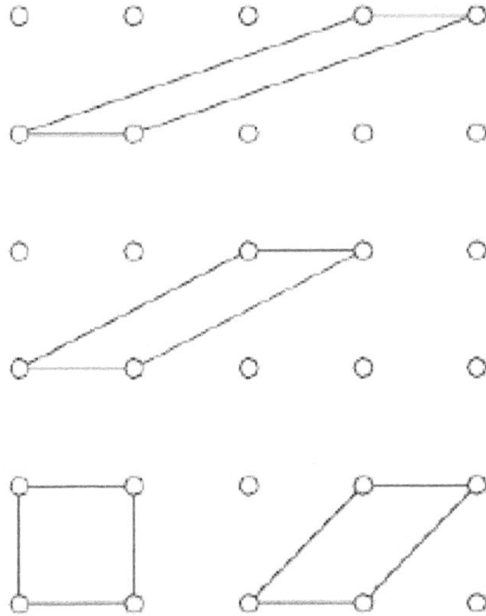

*If we lived in a simulated universe and reached the level of
maximum detail possible in the simulation, we would notice
that the objects would have dimensions equal to the exact multiples
of the minimum possible spacing in the grid. He would not be
possible to have objects or spaces of dimensions more
small or intermediate.*

The universe would therefore be a great memory foam
in which our intention records what we are going to
realize. When I decide for example to sit down, I indirectly give
the information to my energetic body to register movement in
space-time in the form of a physical body. To to ensure the
consistency of the material reiteration, the universe, like us,
functions mainly by habit, our unconscious functioning
representing 96% of brain calculation and based on diagrams
execution specific to each species.
Just as our little conscious potential we allows you to evolve and
make choices, the universe would have also a learning
capacity; that's what we calls morphogenetic fields or morphic

resonance (thesis of Dr Rupert Sheldrake). Through therefore, we have all the ingredients for build reality:
• A support for mass realization (the universe holographic) - the physical body.
• A morphic resonance that reproduces the electromagnetic foam habits - motor and reflex patterns.
• A power of intention which allows the recording movement in space-time – the voluntary functioning and consciousness physics of our body.

9.6 / The universe is not real

One of the great mysteries of modern physics is the reason why antimatter did not destroy the universe at the beginning of time. Physicists assume that there must be a difference between matter and antimatter apart from the electric charge. The physicists at CERN in Switzerland have made the most precise never realized of the magnetic moment of an anti-proton and found it to be exactly the same as that of the proton but with the opposite sign. *"All our observations find complete symmetry between the matter and antimatter, which is why the universe should not really exist,"* says Christian Smorra, physicist at the Baryon-Antibaryon Symmetry collaboration CERN Experiment (BASE). *"An asymmetry must exist somewhere here, but we don't understand everything just not where the difference lies."* Antimatter is notoriously unstable and any contact with regular matter should annihilate it in an explosion pure energy. The Standard Model predicts that the Big Bang should have produced equal amounts of matter and antimatter but it is an oxidizing mixture which would be wiped out, leaving nothing behind to make galaxies, planets or people.

Last year, scientists from the experiment ALPHA (Antihydrogen Laser PHysics Apparatus) from CERN probed for the first time an atom of anti- hydrogen with light, yet finding no difference compared to a hydrogen atom.

A property was only known with moderate accuracy in relation to others - the magnetic moment of the antiproton. Ten years ago Stefan Ulmer and his team at BASE collaboration have given themselves the mission to try to measure it.

They first had to develop a way to measure directly the magnetic moment of the regular proton. They did so by trapping individual protons in a magnetic field and causing jumps quantum in its spin using another field magnetic. Then they had to take the same measurement on antiprotons.

To do this, the team used the coldest antimatter and the most durable ever. After creating the antiprotons in 2015, the team was able to store them for more than a year in a special room of the size and shape of a box of Pringles. Since no physical container can contain antimatter, physicists use fields magnetic and electric to contain the material in devices called Penning traps. Using a combination of two traps, the BASE team created the most perfect antimatter chamber ever retaining antiprotons for 405 days. This storage stable allowed them to perform their current measurement magnetic on antiprotons. The result gave a value for the antiproton magnetic moment of - 2.7928473441 μ N. (μ N is a constant called the nuclear magneton.) Apart from the minus sign, it is identical to the previous measurement for the proton.

The new measurement is precise to nine digits significant, the equivalent of the circumference measurement of the Earth to a few centimeters, and 350 times more specifies that any previous measurement.

9.7 / The time space contains a code of error correction

In the theory of the holographic principle, the universe, the fabric of space and time emerge from a network of particles quantum. Physicists have discovered that it works according to a principle called quantum error correction.

Theoretical physicist Dr. James Gates, who works on string theory for many years, seems to have discovered in string theory, a computer code currently used in the computer search engine technology! This code, known as Block Linear Self Dual Error Correcting Code, is essential for the smooth transfer of computer language and data. He examines the code sent, then compares and measures it against what has been sent and then make the necessary adjustments for the outgoing information to be correct. The Dr James Gates, Jr., professor of physics at the University of Maryland, received his doctorate from MIT and his thesis on supersymmetry was the first ever performed at MIT. He was elected to the National Academy of Sciences in 2013. Its discovery gives credibility serious about the simulated universe hypothesis.

10 / The fascinating similarity between the network neuronal and dark matter

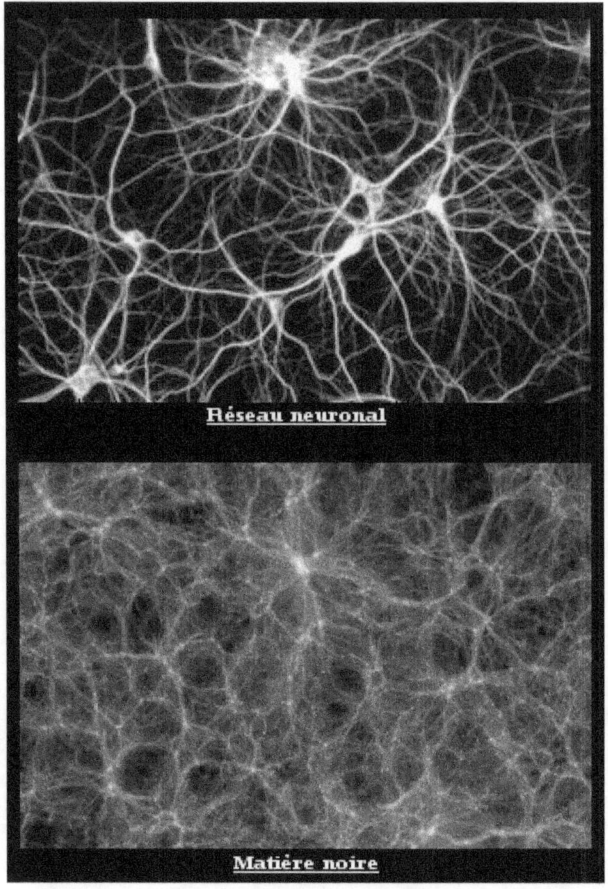

Brain VS black matter

Christof Koch, a leading researcher on the subject consciousness and the human brain, says that this last is the most complex object in the known universe. In effect, with a hundred billion neurons and a hundred thousand billion connections, the brain is an object of dizzying complexity.
But there are many other complicated objects in the universe as for example the galaxies which can be group into clusters, super-aggregates and filaments. The boundary between these

structures and neighboring areas empty space called cosmic voids can be extremely sophisticated.

An astrophysicist and a neuroscientist united our forces to compare the complexity of networks of galaxies and neural networks so quantitative. The first results of their research are truly surprising: the complexities of the brain and of the cosmic lattice are really similar, but so are their structures. The universe can be self-similar on scales that differ in size by a factor a billion billion billion. If the cosmic web is at least as complex as any of its constituent parts, researchers could naively conclude that it must be at least as complex than the brain.

The total number of neurons in the human brain is lies in the same direction as the number of galaxies in the observable universe. But the concept of emergence makes comparison possible. Many phenomena natural are not equally complex at all scales. On smaller scales, with material locked in stars and clouds of matter black, this structure is lost. The universe contains many systems nested in others with little or no interaction at different scales. This segregation at scale allows the study of physical phenomena as they emerge on their own natural scales.

Ordinary matter and dark matter condense into string-like filaments, and clusters of galaxies are formed at the intersections of the filaments, leaving most of the volume remaining practically empty. Cortical gray matter representing more than 80% of brain mass contains about 6 billion neurons (19% of brain neurons) and nearly 9 billions of non-neuronal cells. The cerebellum has around 69 billion neurons (80.2% of brain neurons) and about 16 billion of non-neuronal cells.

It is interesting to note that the total number of neurons in the human brain are located in the same sense that the number of galaxies in the universe observable. In the figure below, we observe a simulated distribution of cosmic matter in a slice 1 billion light years in diameter (established system), with a real image of a slice of 4 micrometers (μm) thick through the cerebellum human. Is the apparent similarity simply the human tendency to perceive meaningful patterns in random data? No, statistical analysis shows that these systems do indeed present quantitative similarities. Researchers use regularly a technique called spectrum analysis power to study large-scale distribution galaxies. The power spectrum of an image measures the strength of structural fluctuations belonging to at a specific spatial scale and tells us how many high and low frequency notes make up the particular spatial melody of each image.

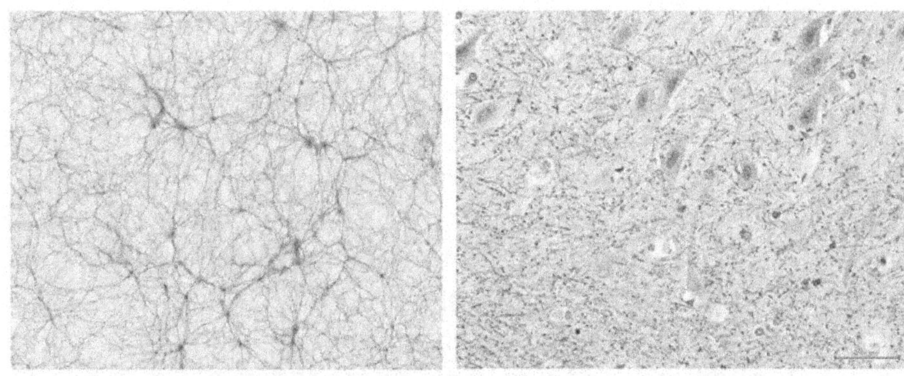

Simulated matter distribution of the cosmic web (left) vs the observed distribution of neuronal bodies in the cerebellum (at right). Neural bodies were stained with the antibody monoclonal clone 2F11 against neurofilaments

An astonishing message emerges from the graphic of the power spectrum in the figure above: The relative distribution of fluctuations in the two networks is remarkably similar, on several orders of magnitude. The distribution of fluctuations in cerebellum at scales of 0.1 to 1 mm is reminiscent of the distribution of galaxies over hundreds of billions light years (established cosmological model). To the most small scales available for observation microscopic (about 10 μm), it is the morphology of the cortex that most closely matches that of galaxies.

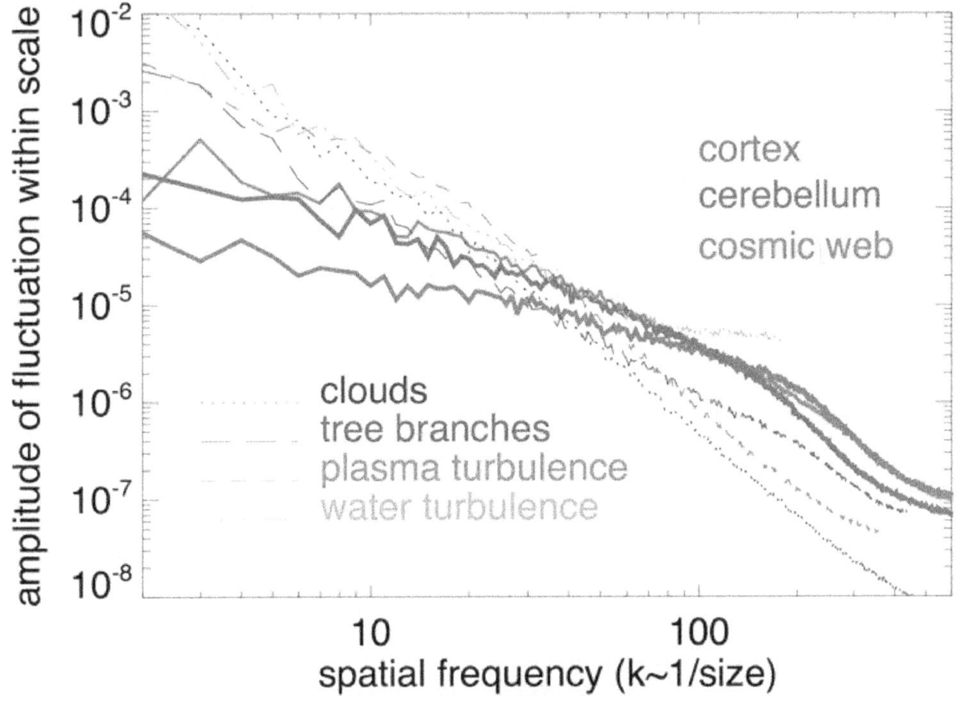

Distribution of fluctuations according to the spatial scale for same cards of the 1st figure. For comparison, the density spectral power of clouds, tree branches and the plasma and water turbulence is also indicated

Based on the latest connectivity analysis of the brain network, independent studies have concluded than the total memory capacity of the human brain adult should be about 2.5 petabytes, not far from the range of 1 to 10 petabytes estimated for the cosmic web!
This similarity in memory capacity means that all the body of information stored in a human brain can also be coded in the distribution of galaxies in our universe, or, on the contrary, that a computing device with the memory capacity of human brain can reproduce displayed complexity by the universe at its greatest scales.

11 / Consciousness and the brain

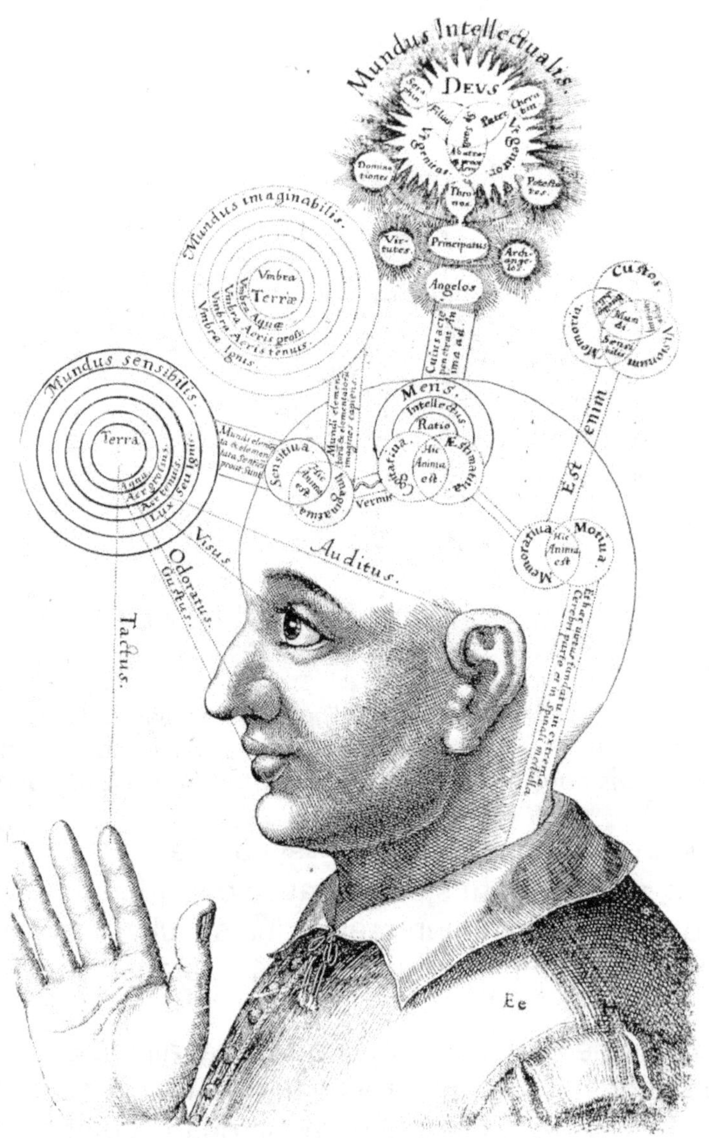

11.1 / The brain

The brain is:
CPU : dual-core multimodal processor with control unit
integrated graphics processing (GPU) and monitor

Neural system : 100 billion neurons with 100 trillions of synapses
Memory : 250 million gigabytes - approximately 400 video stream lifetimes
Eye display resolution : 576 megapixels, approximately 24K
Frame rate : ~ 150 fps
Clock frequency : <1 kHz
Power consumption : ~ 20 watts

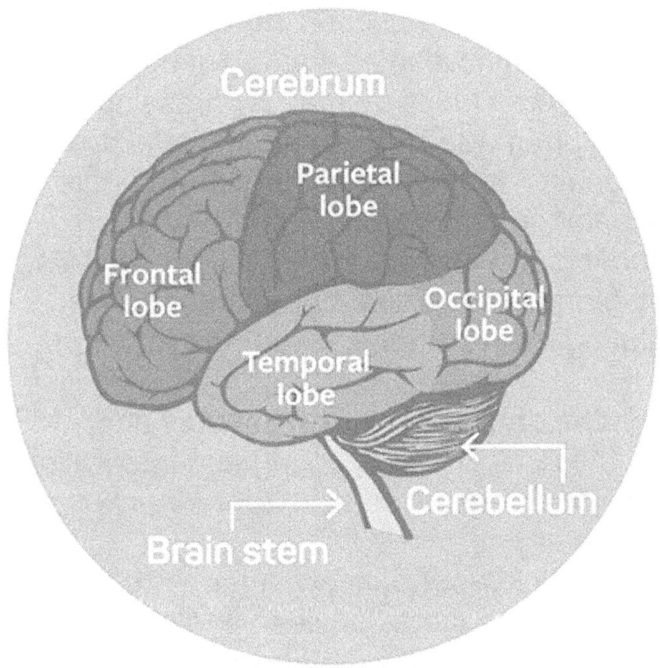

Going up the central nervous system, there is the trunk cerebral which connects the spinal cord to the thalamus, connecting the body to the brain for the treatment of data and feedback control. The brainstem is also responsible for the regulation of several autonomous functions, controlling heart rate, blood pressure, breathing, swallowing and vigilance. The cerebellum is suspended at the back of the brainstem, which is a rather intriguing structure. Her main function seems to be the motor control of precision that encompasses a wide range of applications, such as the joints of the thumb and the movements of the mandible. If the cerebellum does not initiate movement, it contributes to coordination, precision and timing accurate of it. He is responsible for learning engine, adapting and fine-tuning its programs through a process of trial and error.

The cerebellum is made up of a large number of modules more or less independent, all with the same structure geometrically regular internal, performing the same calculation. The nervous system in need of a response fast and clear at the entrance, the cerebellum is able to reduce noise by eliminating the possibility of feedback created by recurrent neural connections. Of studies have shown that the cerebellum is also activated during language, attention and imaging tasks mental. By pressing inwards, what the cerebellum made to movement, it also does to the intellect, to the personality and emotional processing. It is therefore the center of balance, our reptilian brain.

The hypothalamus connects the nervous system to the system endocrine and pituitary gland is essentially a plant hormones under its control, regulating metabolism, growth, reproduction, sleep, mood ...

As for the thalamus, it is the data relay station sensory and motor. It pre-processes the data by transforming and integrating them, to send them to cortical regions suitable for further calculation.

The thalamus is therefore the keeper of knowledge. With using the pineal gland, it also regulates states of sleep and waking consciousness. When the data is transmitted between the thalamus and the neocortex for cognition, they also travel to through the limbic system, the main functions of which are motivation, emotion, learning and memory. He performs emotional processing of order inferior of the inputs of the sensory systems, and it is the seat of our emotional lives. The amygdala, it, takes care of memory processing, taking decision and emotional responses like fear, anxiety and aggression. More exciting stimuli increase the activity of the amygdala, and the level of this activity correlates with memory retention.

The hippocampus is primarily responsible for the memory consolidation, transfer of content from short-term memory and working to storage at long term. It also performs the function of memory spatial which allows us to find our way around.
The limbic system provides an algorithm learning through emotional reinforcement. It's a holographic world mind map with loop feedback for our actions, publicized like them are created by feelings, thus coloring memories that we store and collect.
The neocortex is responsible for brain functions higher order such as sensory perception, cognition, the generation of motor commands, the spatial reasoning and language. He distributes parallel processing of perception, thought and action in separate modules called lobes, each responsible for a particular dimension of experience.

There are 4 lobes of the brain: occipital, temporal, parietal and frontal

The occipital lobe is the visual processing center of the brain. He deconstructs the visual signal of the world by parts and projects it through the prism of the concentration and attention. It creates our perception of three-dimensional space. The temporal lobe is the region where the understanding of sound, language and speech is processed. The parietal lobe plays an important role in the integration of sensory information from the body, with knowledge of numbers and their relationships, and in handling objects. This is the main area of bodily and spatial awareness. She holds the power of association.
The somatosensory cortex is a region of the lobe parietal which receives and processes sensory inputs from all the body. Finally, the motor cortex, part of the lobe front end, manages the planning, control and execution of voluntary movements. The frontal lobe is the part of the brain that controls cognitive skills important such as emotional expression, problem solving, memory, language, judgment and sexual behavior. It's the

personality control panel and our ability to communicate. The frontal lobe is seen as the seat of self-awareness. Each layer consists of a characteristic distribution of neurons and connections to other cortical regions and subcorticals, including the thalamus. Each of the brain lobes are made up of columns of cells nerve layers and these nerve calls specialized are neurons. White matter contains few cell bodies and consists of mainly long-range axons. Grey matter, on the other hand, contains a large number of bodies neuronal cells but few axons. Cells glia are non-neuronal cells in the central nervous system, which means that they do not produce of electrical impulses, but play an important rôle in maintaining homeostasis by providing protection to neurons. Glial cells are essential for the functioning of neurotransmitters.

Gray matter and glial cells, taken together, are neural dark matter, the biochemical grid underlying supporting the electrical oscillation in the brain, weakly interacting with neurons in the white matter.

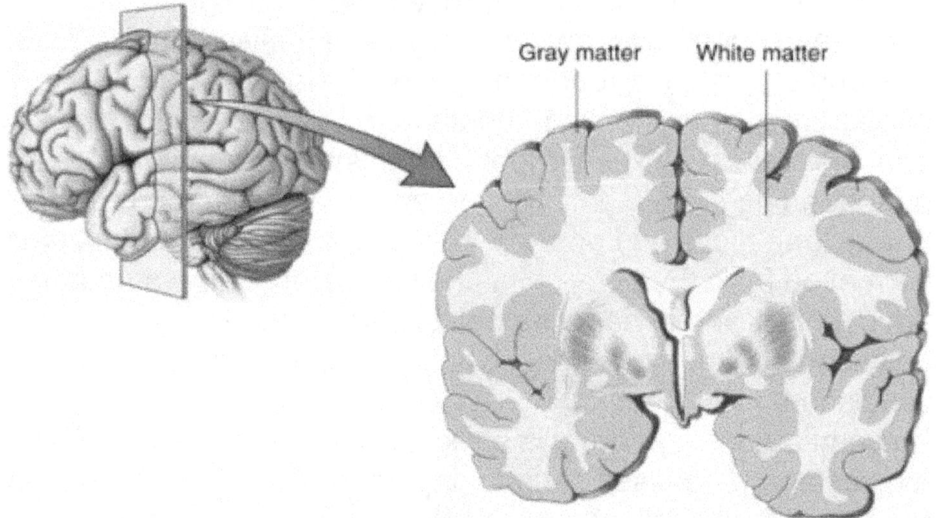

The claustrum is a thin bilateral surface that connects the cortical regions to the thalamus. It is the most densely connected in the brain allowing the integration of various cortical inputs (color, sound and touch) in an experience rather than singular events.

And because of its location deep inside brain, the study of the claustrum remains difficult.

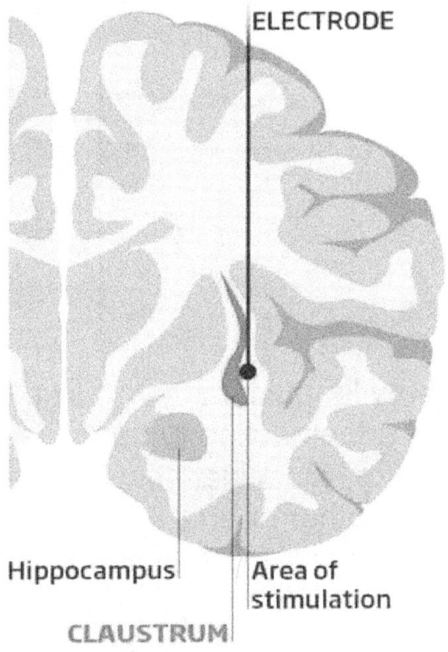

However scientists believe it plays a key role
in awareness and attention. Then in the center of brain,
supplying it with nutrients and eliminating its toxins are found in
the ventricles. It is a space dark filled with cerebrospinal fluid
(CSF) essential for all brain functions. When we sleep, the brain
flushes the ventricles with CSF and removes toxins, providing
the brain with new resources and preparing it for another wave of
sensorium on waking.

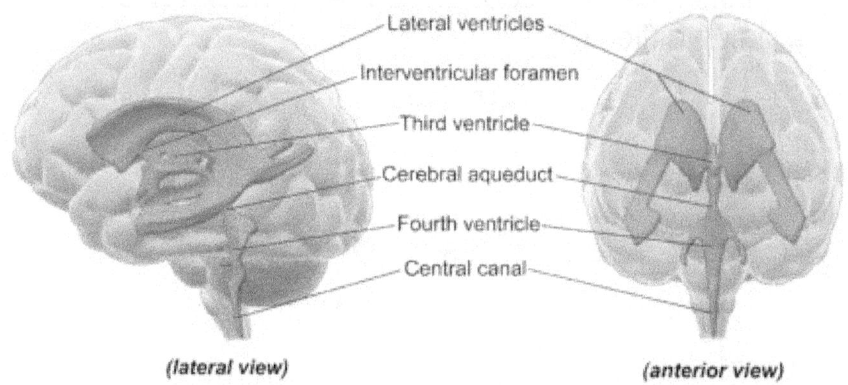

Staying awake over a very long time is very toxic.
When a neuron is triggered, an action potential is
moves along its axon to the slits synaptic, where it forms a
junction with the dendrites other neurons. The action potential
causes the release of neurotransmitters that bind to receptors on
the postsynaptic neuron, opening the gates to the passage of
electrically charged ions, and creating another action potential
that spreads along of the axon. The action potentials can be
enough powerful if there is sufficient time to that interesting
structures evolve for their transmission. Only the organizations
most suited to their environment are able to adapt enough to
ensure their survival and prosperity.
As such, they are the catalyst for neural computation,
a process sometimes resulting in the formation of a
thought. By controlling the flow of information in the
brain with small data packets, neurotransmitters optimize the
experience. Once that the brain is started, information circulates
in the nervous system in feedback loops by processing sensory
data and calculating a response to these.
From an electromagnetic point of view, we can
visualize this as oscillation waves neuronal, distributed over
several bands of frequencies, calculating different modes of
consciousness, that is different descriptions of the same
phenomena.

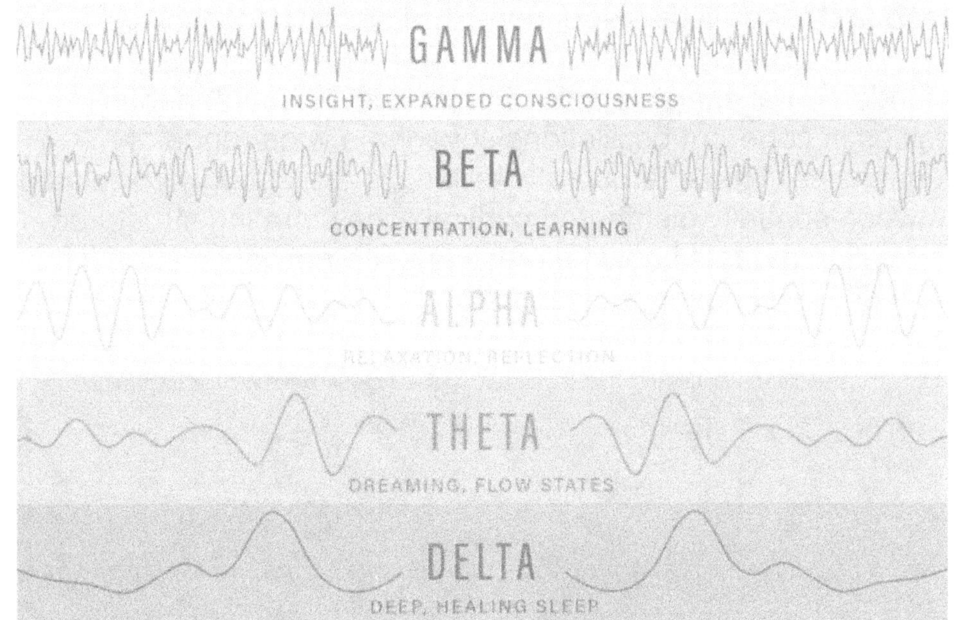

The code of consciousness is carried on 6 bands of frequencies:

Epsilon (90+ Hz)
Reality modulation and gain control

Gamma (30 to 90 Hz)
Higher cognitive function and motor control

Beta (13 to 30 Hz)
Normal wakefulness, concentration, physical senses

Alpha (8 to 12 Hz)
Relaxed, light meditation, creative, learning, aware

Theta (4–8 Hz)
Light sleep, deep mediation, creativity, intuition, reminder, fantasy

Delta (0.5 to 4 Hz)

Deep and dreamless sleep, non-REM sleep, unconscious

A fundamental characteristic of brain activity spontaneous resides in coherent oscillations covering a wide range of frequencies. These temporal oscillations are strongly correlated between spatially distributed cortical areas forming structured correlation models known as resting state networks, even if the brain does not never really rests.

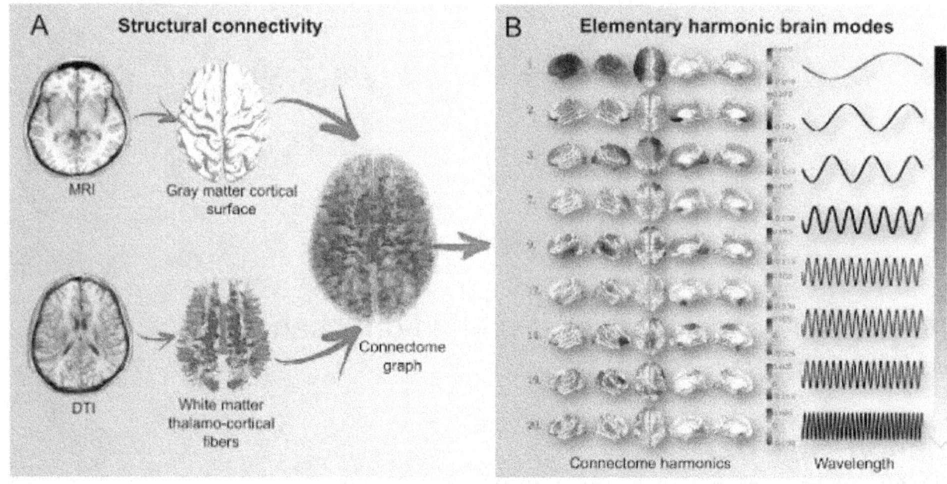

The structure of the brain naturally gives birth to distributed clusters of neurons that resonate and are triggered together according to the frequency of oscillation. Provide targets for a calculation distributed, multidimensional and in real time. Each hemisphere controls the other half of the body and each from them has a distinct point of view. Hemisphere left analyzes things, manages logic and language, the right hemisphere is concerned with the sensation of things and creativity. And connecting the two hemispheres is the corpus callosum, a large and thick, consisting of a flat bundle of fibers, under the cerebral cortex in the brain. The function of the corpus callosum is perhaps best illustrated when one considers his absence. Once the brain is right and left separated, each hemisphere will have its own perception, its own concepts and its

own impulses to act. The overlap of perspectives in our brain,
patterns
interference from different points of view, create
the experience of depth.

11.2 / Where is the consciousness?

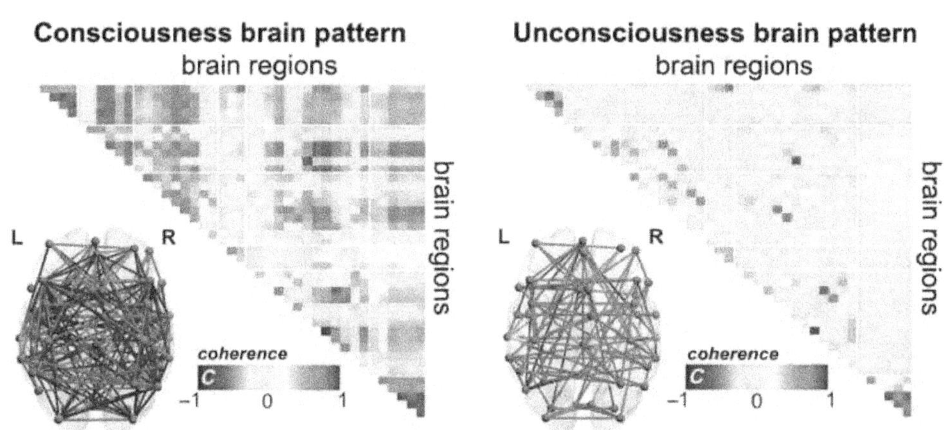

*The difference between the brain activity of a conscious brain
(left) and an unconscious brain (right). © E. Tagliazucchi & A.
Demertzi*

Where to place consciousness in this inextricable whole of
interconnected circuits of the brain? The scientists, who hoped to
locate and circumscribe its source, know now that this quest
does not make sense. Instead of attempt to elucidate the
problem directly, the researchers in cognitive neuroscience
prefer to discover the brain events that accompany
consciousness.
They observe what happens in the brain during a particular
conscious experience, compare it to unconscious situations, and
determine what are differences in neural processes. They
can thus build bridges between subjective states (feelings) and
mental states (physical states, measurable and observable). The
first results have shown that consciousness is not localizable, but

that in addition it turns out to be almost useless to our activities cerebral! Most of the visual data, hearing, and touch is processed by the brain without even let him realize it. In the early 1980s, the English psychologist Anthony Marcel highlighted that one can have understood the meaning of a word without even to have been aware of having seen it!

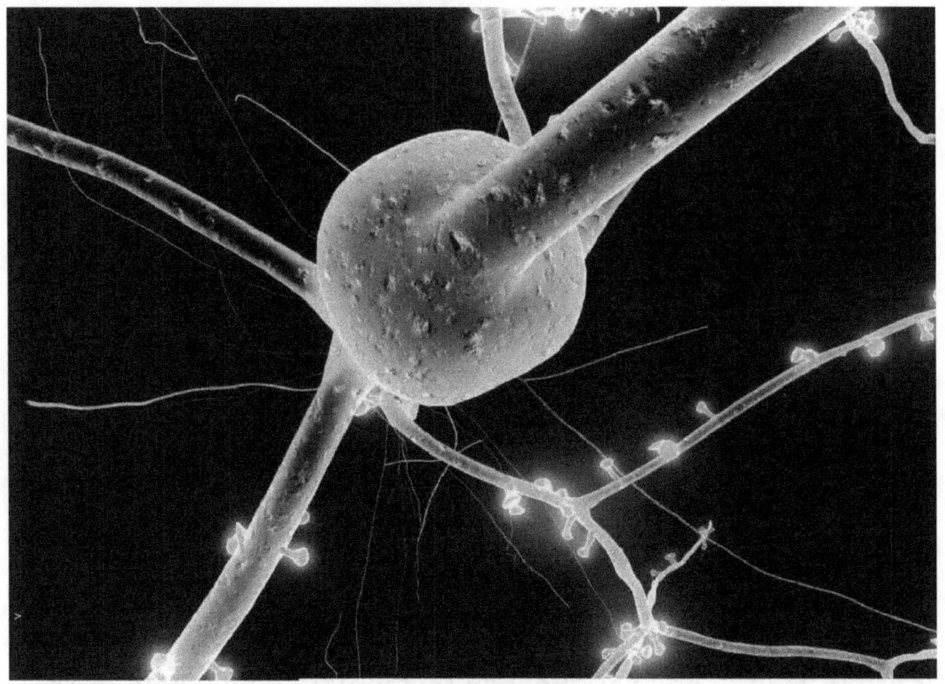

Individually, cells like neurons are devoid of "spirit". However, their assembly brings out the consciousness.

So the curtain falls on this simplistic vision that assimilates our consciousness to a place in our brain where information would be collected and from from which decisions would be made.

Consciousness during lucid dreaming

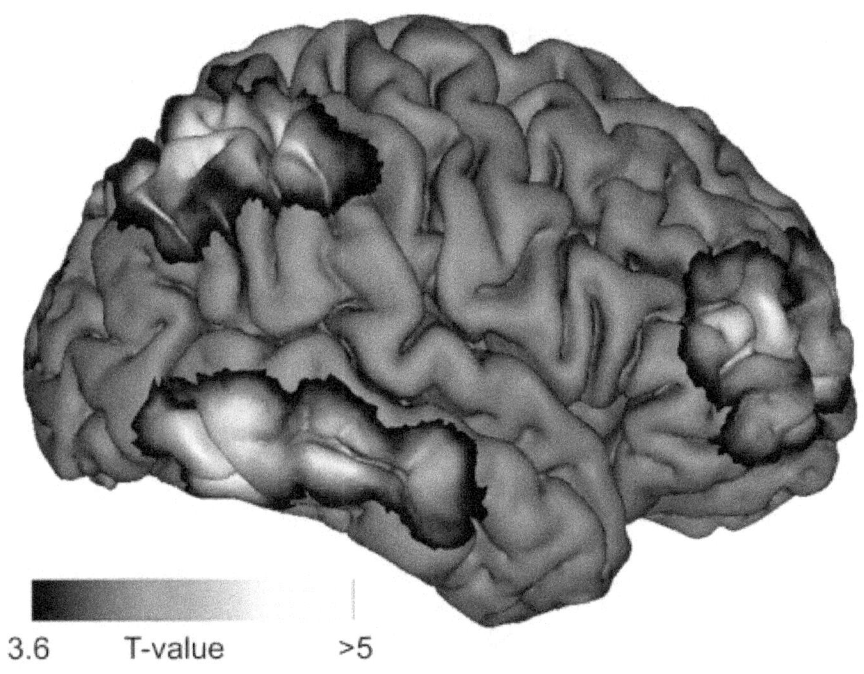

*Specific areas of the brain are activated more strongly
during lucid dreaming than in normal dreaming*

Neuroscientists from the Max Planck Institute in Munich and
human cognitive sciences and brain in Leipzig and Charity in
Berlin have identified a specific cortical network associated with
self-awareness.
They used EEG and fMRI brain imaging to study lucid dreamers,
who have access to their memories during the dream and are
aware of themselves, while remaining in a dream state and
without wake.
Researchers have found neuronal activations in a specific
network which is normally disabled during REM sleep, including
these areas:
• Dorsolateral prefrontal cortex (associated with an evaluation
self-focused metacognitive)
• Dorsolateral prefrontal cortex in association with parietal lobules
(may reflect demands for working memory)

• Bilateral frontopolar areas (linked to the treatment of internal states, for example, the evaluation of one's own thoughts and feelings)
• Precuneus (involved in self-referential treatment, such as first person perspective)
• Bilateral cuneus and occipitotemporal cortex (active in conscious awareness in visual perception)
The study was limited to only four subjects who were highly skilled lucid dreamers. Only one of them became lucid twice under EEG conditions / Simultaneous fMRI, making our data a study of cases. "The researchers also indicated that part of the observed activation may come from the task of eye and hand signaling performed during the lucid dreaming process.

11.3 / Consciousness under LSD

Scientists have observed an increase supported by the diversity of neural signals, a measure of the complexity of brain activity, people under the influence of psychedelic drugs, compared to when they were in a state of Eve. It has been shown that people who are awake have a more diverse neuronal activity in using this scale than those who are asleep.
However, this is the first study to show a diversity of brain signals greater than the value of base, higher than in someone who is simply awake and aware. Previous studies have had tendency to focus on states of consciousness lowered, such as sleep, anesthesia or the so-called condition vegetative.
Professor Anil Seth, Co-Director of the Sackler Center for Consciousness Science at the University of Sussex, declared:
" *This discovery shows that the brain on psychedelics behave very differently from the normal. During the psychedelic state, the activity electrical power of the brain is less predictable than during normal conscious wakefulness - as measured by global signal diversity* " .

For the study, Michael Schartner, Adam Barrett and the
Professor Seth of the Sackler Center re-analyzed
data previously collected by Imperial College of London and
Cardiff University in which healthy volunteers received one of
three drugs known to induce a psychedelic state:
psilocybin, ketamine and LSD. Using the brain imaging
technology, they measured the tiny magnetic fields produced in
the brain and found that for all three drugs, this measure of the
conscious level was higher by reliably. The results could help
shed light the growing discussions about the use of carefully
controlled medical treatment of these drugs, for example in the
treatment of severe depression.
In addition to helping inform medical applications possible, the
study adds to an understanding increasing scientific knowledge
of how the level of consciousness and conscious content are
related to each other.

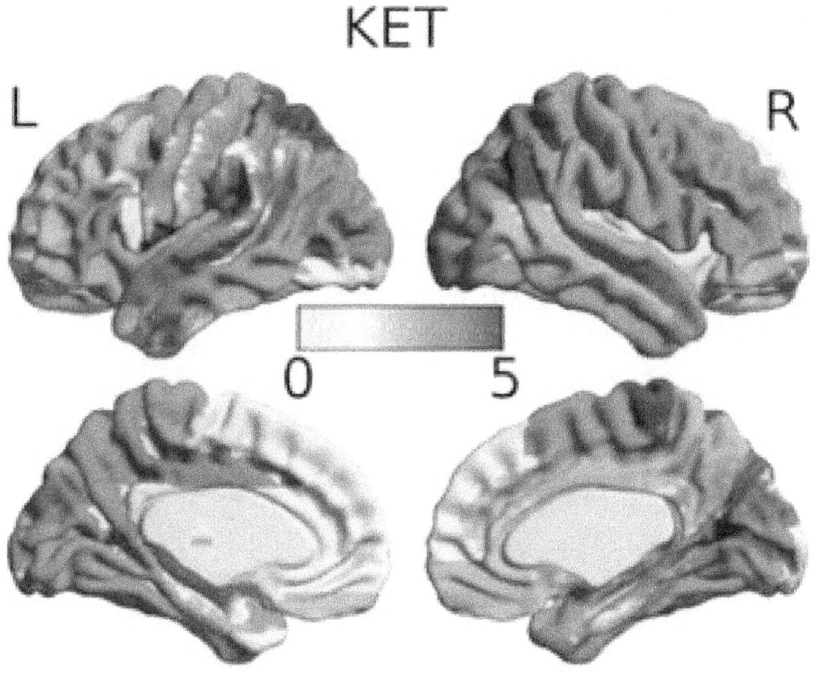

The brain on psychedelic drugs, stamen or psilocybin, has a higher level of neuronal activity (red) than normal state

Serotonin 5HT2A

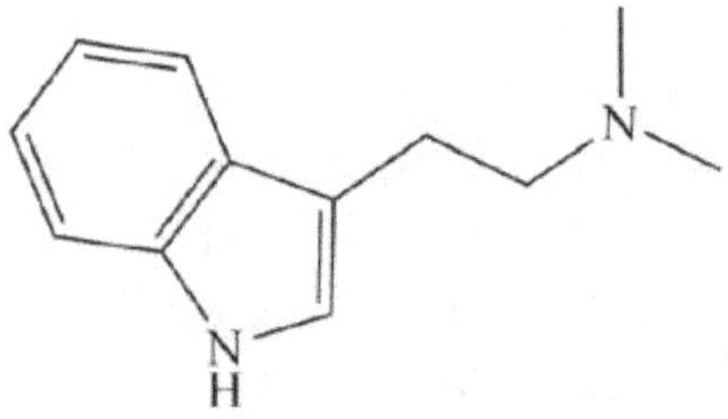

DMT

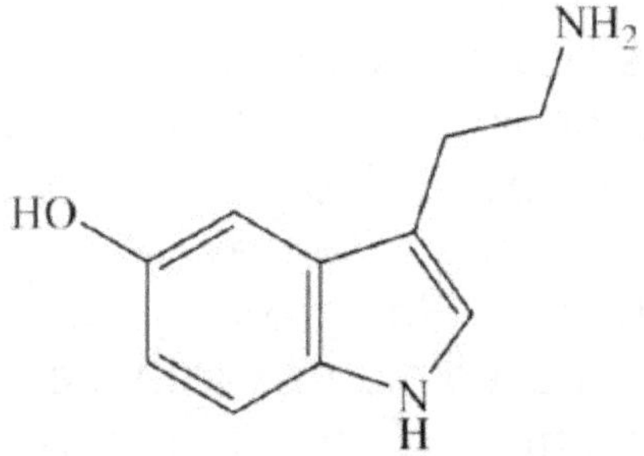

Serotonin

Under normal circumstances, levels of serotonin (and probably DMT (dimethyltryptamine)) in the brain are precisely tuned to reduce the wave function of the hologram in what we see around us, balancing survival needs against those of imagination. With the contribution of psychedelic molecules - serotonin 5HT2A sub-receptor agonists - we are subject to a spectacular increase neural shots all over the brain. In addition, they help to prevent the reuptake of serotonin once it is has been secreted, which further increases the rate of triggering of neurons.
Basically we flood the brain with neurotransmitters and signal overload independent of external stimuli.

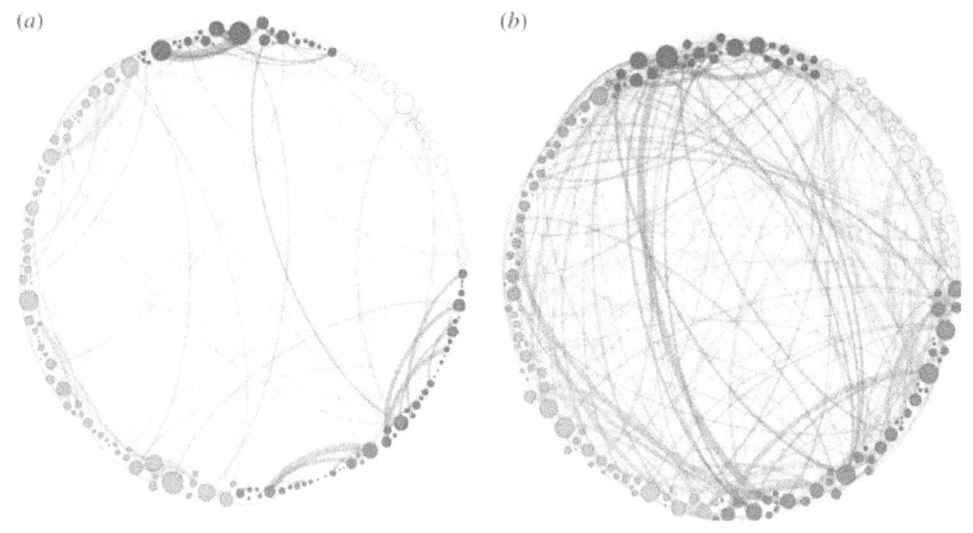

The effects of psilocybin on the interconnectivity of regions cerebral

A consequence of the hyperactivity of neurons and
their increased trigger rate is a hyper-connectivity of a large
number of brain regions to increased power levels.
A large number of connections are made throughout
the cortex which otherwise would not have been activated.

Hallucinogenic mushrooms Psilocybe semilanceata

Another consequence of the change in the rate of
neural triggering is a change in the brain waves themselves. As
the power is increased and redistributed in a network much
larger neural interference patterns and brain waves are
dramatically altered.

As proposed, brain waves are the carrier frequencies of
consciousness, calculating its modes. And the substantial
changes in their power should be correlated with changes in
subjective experience. And with DMT, we find that this is clearly
the case. Changes of consciousness are due to a change in the
curvature of the spatio-temporal geometry of the simulation. The
concrete hypothesis is that the network of measures subjective
of the distances that we experience on DMT (derived from
relationships between objects phenomenal that one experiences
in this state) has a overall geometry that can be accurately
described as hyperbolic. The spatio-temporal geometry of the
simulation is inherently hyperbolic. Our inner world becomes
bigger than it is possible to to adapt in an experiential field with a
space phenomenal Euclidean 3D. It follows that spaces,
phenomenal surfaces and objects acquire a mean negative
curvature. Under the influence of psychedelic, the simulation
acquires a curvature negative when overloaded with information.

The geometric levels of a journey psychedelic
1. Visual noise

Geometry is perceived as visual noise or static, combined with
areas of stray light and dark red that appear under the eyelids.

2. Movement and color

The effect can be described as the appearance of regions not
structured with sudden lightning and clouds of color. These are
usually called phosphenes and can often be felt in a sober state
in rubbing or applying pressure to or near the
eyes closed.

3. Partially defined geometry

Relatively complex shapes and patterns with a vague structure begin to form. These patterns remain strictly two dimensional. The geometry is thin, small and enlarged with a palette of colors dark which is generally limited to a few shades different, such as blacks, reds and purples dark.
They are displayed in front of the visual fields of the eyes open and closed through a flat geometry veil but they are much more detailed with the eyes closed or in dark environments.

4. Fully defined geometry

The detail in which the geometry is displayed becomes deeply complex and totally structured, but still remains strictly two-dimensional.
The geometry becomes larger and extremely intricate in detail with a color palette almost unlimited in its possibilities.
They are displayed both on the open visual field and closed my eyes through a flat veil of geometry which floats directly in front of a person's eyes, remaining much more detailed with eyes closed or in dark environments.

5. Three-dimensional geometry

At this level, the geometry has become entirely three-dimensional in its shape and position in the visual field, adding a new layer of visual complexity and leaves the geometry spread over the surfaces and objects in a person's environment instead of just being displayed through a veil basic and flat in front of its visual field.

6. Partially replace visual perception

The geometry has become so intense, vivid and shining that she began to block and replace the world outside.
Visual perception of a person's environment begins to be replaced by geometry.
With objects and settings transforming into complex geometric structures. From this level geometry, it is even possible to see a geometry perceived as four-dimensional or created from new geometric principles, no Euclidean or absurd, although this is more common at higher levels.

7. Completely replace visual perception

The geometry continues to become brighter and brighter, and begins to replace the outside world. The meaning of eyesight becomes completely impaired. This creates the perception that we are no longer in the environment exterior, but in another reality of forms extremely complex geometric shapes, such as from another world.

8A. Perceived exposure to the concept network semantics

A level 8an experience can be described as the feeling of being exposed to a mass apparently infinite geometry composed entirely of understandable representations, which are perceived for simultaneously transmit each concept, memory, neurological process and structure stored internally in the spirit.
These sensations are experienced as conveying a
equal amount of information inherently understandable than those which are also felt through a person's vision.

8B. Perceived exposure to the inner mechanics of the consciousness

The level 8B experience can be described as the feeling of being exposed to a mass of geometry composed entirely of geometric representations intrinsically readable which subjectively give the impression that they convey the internal mechanics which makes up all the sub-neurological processes underlying.

During this experience, the organization, the structure and the programming behind conscious experience of a person are perceived to be conceptually understood.

This whole experience is seen at through visual geometric data naturally understood and is also felt physically at an incomprehensible level of detail by supporting cognitive sensations and complex touch screens.

Testimonials from "experimenters"

I wanted to know what hallucinogenic substances visually provoked in those who took it.

So I searched forums and what I could there finding did not disappoint me, because the majority of witnesses speak of a "grid" which would correspond perfectly to the energy grid surrounding the Earth and connecting the "all".

Here is a collection of the different testimonies:

"The only time I remember seeing a gate, it was under mushrooms. It was as if it was the underlying structure of this apparent reality but I don't really know. It was really simple without any detail, just a simple grid with a perspective of apparent depth. It reminded me of a holodeck, it remained motionless while the body was returning. It looked a lot like this but without all the extra stuff:

"When I was in college I tried mushrooms for the first time and at the top the sky looked checkered, as if I were looking through a fence at chain links, and me and the other two people with whom I was saw the orange, in a spiral lift it up sky like a mini hurricane made from fires fireworks. And then we all saw half the city that we were overlooking (we walked to mountain top, good time) completely darken for about 5 seconds.
The next day we asked around and we found no reason why half of the city would have vanished. Nobody saw anything.
Definitely one of the coolest trips. "

"On DMT. I basically enter the code of the universe
(I'm sure my brain does all of this but it seems so real). It's almost like I'm an entity in the coding of a simulation. But my body is not there, just my conscience. Very difficult to put in words visual / mental experience. "

" I saw the gate. I saw it the first time I took 4-Aco-DMT.
It looked like a green and yellow energy grid perfectly symmetrical with energy points at its intersection points and connections that could not be seen only when I focused on it or when I saw him out of the corner of my eye. Visibility of intersections was quite important. The space between each point of intersection was, by memory, as wide as my forearm and

some of the energy connections between intersections seemed to be blurring.

Which was strange and made me believe it was a further proof of the reality of taking substances based on tryptamine, it was the fact that the two friends i was partying with also have it seen. We confirmed this shared experience in autonomously pointing various points intersection - the fear and joy we had in the face ofthis mystery were incredible. We thought it was space / time itself. We haven't seen it since. I then told another friend who explained to me that he had already told me, the story of the first time that he took LSD, when he saw "force fields" energy ". I questioned this friend more and also another friend who he was with and aligning their experiences with mine, we have confirmed that it had to be the same grid - THE grid. I think the metaphysics is the opposite of current physics -
two ends of the same stick. Some of the answers that physicists are looking for with machines like the LHC can already be found and seen during taking substances that give rise to experiences metaphysics. After much thought and search on the grid, I believe that is not the fabric of space / time but the Higgs field, the invisible energy that permeates the universe. "

11.4 / The quantum brain

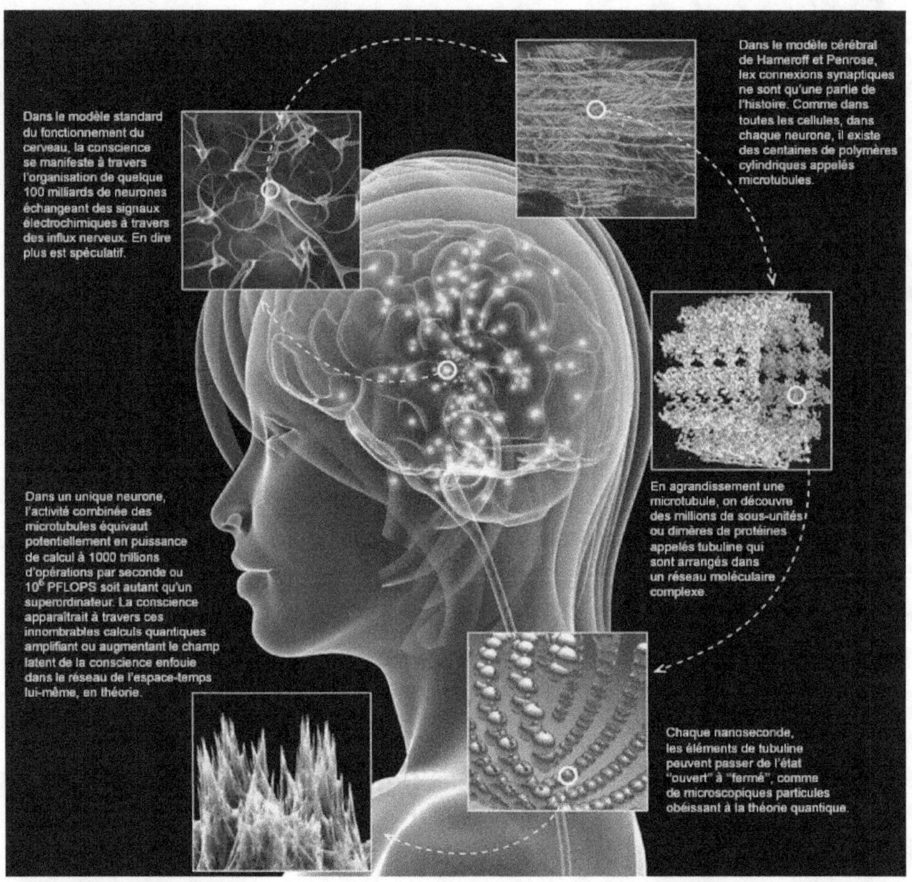

The potential characteristics of quantum computing could explain the enigmatic aspects of consciousness. The Penrose-Hameroff model (reduction orchestrated objective: Orch OR) suggests that the quantum superposition and a form of quantum computation occur in microtubules, networks cylindrical cells of cell cytoskeleton proteins in neurons. Microtubules couple and regulate synaptic functions at the neuronal level, and may to be perfect quantum computers because of the dynamic network structure of subunit states at the quantum level and the intermittent isolation of environmental interactions. Objective reduction orchestrated (Orch OR) is a theory which proposes that the consciousness consists of a sequence of events discrete, each being a moment of objective reduction (OR) of a quantum state, where it is assumed that these states quantum exist as parts of a computation quantum carried out mainly in microtubules neuronal. Such RO events should be orchestrated appropriately (Orch OR), to that a real consciousness arises. OR itself is considered ubiquitous in stocks physical, representing the link between the worlds quantum and classical, where quantum superimpositions between pairs of states spontaneously resolve into classical alternatives in a time scale τ, $\sim$ calculated from the amount of mass displacement that there is between the two states.

In our own brain, the OR process that evokes consciousness would be actions that link the biology of brain (quantum calculations in microtubules) at the fine-scale structure of space-time geometry, the most basic level of the universe, where the tiny quantum space-time displacements are considered responsible for the operating room.

The Orch-OR proposal therefore extends to a range considerable number of scientific fields, touching on foundations of general relativity and mechanics quantum, in an unconventional way, in addition to more obviously relevant areas such as neuroscience, cognitive science, biology molecular and philosophy. It is therefore not surprising that Orch OR has been constantly criticized under many angles since its introduction in 1994.

Nevertheless, the Orch OR scheme has so far better Withstood the test of time than most others patterns, and it is particularly distinguished from other proposals by the many ingredients scientifically tested and potentially testable which it depends. The information states of tubulin in the quantum and classical calculus of Orch OR are now correlated with dipoles, rather than with a mechanical conformation, thus avoiding the problems of heat and energy.

The tubulin dipoles involved in the calculation and the entanglement can be electrical (separation London force charges) or magnetic (states and electron spin currents). Conductance improved electronics discovered in single warm-temperature microtubules at specific frequencies of alternating current gigahertz, megahertz and kilohertz strongly supports OR Orch. BC and Orch OR may well be channel-mediated intratubulin quantum aromatic rings, as in photosynthetic proteins, so plausible for quantum computation in microtubules.

Anesthetics bind in these quantum channels tubulin, presumably to disperse dipoles quantum values necessary for consciousness. The process scale invariants (1 / f of fractal type) at the level of neurons and network could perhaps expand down to the intra-neuronal BC in the microtubules. The Orch OR proposal suggests that conscious experience is intrinsically linked to fine-scale structure of the geometry of space-time, and that consciousness could be deeply linked to the functioning of the laws of the universe.

11.5 / How to know if you are not the only one to be aware of the universe?

Philosophers call it the problem of others minds, others the problem of solipsism. Solipsism is an extreme form of skepticism, both insane and irrefutable who maintains that you are the only one conscious existing. The cosmos is born when you are become conscious, and it will disappear when you die. As crazy as this idea is, it is based on a crude fact: each of us is locked in a prison cell impermeable to consciousness subjective. You experience your own mind every waking second, but you cannot infer the existence of other spirits only by means indirect. Other people seem to have conscious perceptions, emotions, memories, intentions, just like you, but you can't not be sure that they are not simulated. You can guess what the world is like but you will never have direct access to its real existence physical in the past (before your creation). The natural selection instilled in us the ability to understand the emotions and intentions of others but we have a counter-tendency to get them wrong each other and fear of being deceived. The problem of solipsism thwarts efforts to explain the consciousness. Pan-psychists maintain that all creatures and even inanimate matter, even a single proton, possess consciousness. Solipsism is a paranoid but understandable response to feelings of loneliness that lurk in all of us. The religion is an answer to the problem of solipsism. Our ancestors imagined a supernatural entity that testifies of our innermost fears and desires. It doesn't matter to how lonely we feel, how much we are far from our fellows, the creator is always there to watch over us.

Love, ideally, gives us the illusion of transcending the problem of solipsism. You feel that you really know someone, from the inside out, and she will knows. In moments of sexual communion ecstatic or worldly conviviality, the barrier between you seem to disappear. Inevitably, however, your lover disappoints you, deceives you, betrays you. For be less dramatically, a slight bio-change cognitive occurs. The problem with solipsism is reappeared, more painful and suffocating than ever. It getting worse. In addition to the problem of other minds, there is the problem of ours. As the psychologist points out Robert Trivers, at least we're wrong too effectively that we deceive others. A corollary of this dark truth is that we know each other even less than the others. According to the Buddhist doctrine of the anatta, the ego does not really exist. When you try to pin your own essence, to seize it, it slips through your fingers. The solipsism is a cave you can't get out of except possibly by pretending that there is not. Maybe the best way to face the problem of solipsism is to imagine a world in which he disappeared.

11.6 / Dreams

Everything we see, hear, touch, taste and feel is information decoded by our consciousness to make our senses 'real'.
The dream world is just another frame of virtual reality in which our consciousness can dwell temporarily. Every time we sleep we dream, even if we forget most of the dreams that we do. In every new dream our mind constantly inventing new scenes. How a brain, residing inside a skull that cannot see never a single photon of light, can it create in darkness a world where everything is illuminated? It's because our mind has the capacity to create whole worlds to starting from scratch - all of this being done through our subconscious. This is how the brain really is powerful.
The landscape of the world of dreams or nightmares and what we can do there, like fly, drown, fall into the void without dying or being hit by a ball without suffering, it is because it has a set rules different from our "reality".
Our everyday world has a set of rules rigid where physical processes do not evolve a lot; if you place your bag on your sideboard, there is chances are he'll be here tomorrow. But in the world of dreams, everything is in motion. Dreams can end as quickly as they started and they do not follow any rules. Another big difference in the dream world is time, which is very different

of the everyday world. It stretches or expands. But why do we dream and what do dreams mean? He seems that dreaming is a way of solving problems and to change our consciousness. In our reality, we face problems permanently. In dreams we are too faced with challenges. And that's the way we them treat that changes the level of learning, such as in life. In the real world, we often take decisions from our intellect, but the world dreams are the reverse.

Whoever runs our simulation, provides the dreams that we are experimenting. Often these dreams put us in situations that test us and we challenge by stimulating our ego and our anxieties. Yes we always make the wrong decisions, we will have to relive the same dream over and over again until we take the good. The dream can come back again to make sure that we are above our fears. Dreams are not there to haunt or scare us, but to help us learn and improve quality of our consciousness.

Most dreams are an extension of what is happening produced in everyday life. And the reverse is also true: the dream can have an impact on our lives.

Currently, therapists help patients to interpret their dreams, through their stories, search for clues, symbols and structures that could fit their life.

Castle's Hall and Van system is the most known to analyze dreams. He codifies the characters, interactions and their effects. Only problem: it is a laborious process which requires manual work.

This is why sleep scientists are looking an algorithmic way that would automate this mission. Researcher Fogli and his team found the solution by analyzing 24,000 dreams, from a huge public database of dream reports, named DreamBank.

" We have designed a tool that automatically assesses dream reports by operationalizing the scale dream analysis widely used by Hall and Van de Castle. We have validated the effectiveness of the tool on handwritten dream reports and tested what is called the assumption of continuity at this unprecedented scale.

These three dimensions are considered to be the most important to help in the interpretation of dreams, because they define the backbone of a dream plot: who was present, what actions were carried out and what emotions were expressed, " the team explains.

When they compare the results of their tool with the handwritten notes of dream reports written by dream experts, the results corresponded to the three quarters of the time. Even if it's not a perfect score, it is a promising way to develop similar technologies in order to unravel the mysteries of search for the dream.

Using the evidence, researchers found a check that supports the assumption of continuity, know that dreams are a continuation of what is lived in everyday life. Always according to researchers, the dreams contained various markers statistics reflecting what dreamers may have lived in real life.

This study will thus facilitate the quantification of the aspects important dreams, and could build technologies that bridge the current gap between life real and dream.

11.7 / An explosive CIA document; "The gateway process"

The Gateway process is a training system designed to improve concentration and consistency of amplitude and frequency of brain wave output between left and right hemispheres in order to modify the awareness and escape the restrictions of time and space.

The Gateway experience was originally created by Robert Monroe, founder of the Monroe Institute (TMI) which is without doubtless the most renowned institute for the studies of conscience since 1971.

The gateway is to experience the aspects not physics of reality and teaches us to what extent our consciousness is really powerful.

To understand the processes involved in achieving these kingdoms, members of the US government looked deeply at the states of consciousness altered. They studied hypnosis, transcendental meditation, biofeedback, hemisync and binaural beats. They also have taken a large part of the information and the experience gained through the MK Ultra project. Once their research completed, a document was created in 1983.

Approved For Release 2003/09/10 : CIA-RDP96-00788R001700210016-5

DEPARTMENT OF THE ARMY
US ARMY OPERATIONAL GROUP
US ARMY INTELLIGENCE AND SECURITY COMMAND
FORT GEORGE G. MEADE, MARYLAND 20755

IAGPC-O

9 June 1983

SUBJECT: Analysis and Assessment of Gateway Process

TO: Commander
US Army Operational Group
Fort Meade, MD 20755

1. You tasked me to provide an assessment of the Gateway Experience in terms of its mechanics and ultimate practicality. As I set out to fulfill that tasking it soon became clear that in order to assess the validity and practicality of the process I needed to do enough supporting research and analysis to fully understand how and why the process works. Frankly, sir, that proved to be an extremely involved and difficult business. Initially, based on conversations with a physician who took the Gateway training with me, I had recourse to the biomedical models developed by Itzhak Bentov to obtain information concerning the physical aspects of the process. Then I found it necessary to delve into various sources for information concerning quantum mechanics in order to be able to describe the nature and functioning of human consciousness. I had to be able to construct a scientifically valid and reasonably lucid model of how consciousness functions under the influence of the brain hemisphere synchronization technique employed by Gateway. Once this was done, the next step involved recourse to theoretical physics in order to explain the character of the time-space dimension and the means

Declassified in 2017, here is the full translation:

SUBJECT: Analysis and evaluation of the process of gateway

A: Commander of the army task group
American Fort Meade, MD 20755

You asked me to provide an assessment of

experience of the bridge in terms of its mechanics
and its ultimate utility. As I proceeded to fill this
task, it quickly became apparent that in order to assess
the validity and practicality of processes whose
I needed to do enough research for defense and analysis to fully
understand how and why the process works. Frankly, Sir, this
has been extremely difficult. Initially based on conversations with
a doctor who followed the `` bridge '' training with me, I used the
biomedical, models developed by Itzhak Bentov
for information about the appearance physics of the process. So
I found it necessary to search different sources for information
concerning quantum mechanics in order to be able to describe
the nature and functioning of consciousness human. I had to be
able to build a model scientifically valid and reasonably lucid of
the how consciousness works under the influence of cerebral
hemisphere synchronization technique used by gateway. Once
done, the next step was to use physics to explain the character
of the spatio-temporal dimension and the means by which the
expanded human consciousness transcends it in the
achievement of Objective Gateway.
Finally, I still found it necessary to use physics
to bring the whole phenomenon of states out of the
body in the language of physical science to remove the
stigmatization of its occult connotations, and put it in a frame of
reference adapted to objective assessment. I started the story in
briefly presenting biomedical lactors fundamentals affecting
related techniques such as that hypnosis, biofeedback and the
transcendental, the meditation so that their goals and their way
of operation can be compared to the reader's mind with the
experience of the catwalk as a sound model underlying
mechanics have been developed. In addition, this introductory
material is useful in supporting conclusions of the document. I
indicate that sometimes these related techniques can provide
points useful input to accelerate movement in the catwalk
experience.

Niels Bohr, the famous physicist, has already answered the complaints from her son about the obtuse nature of some concepts in physics by saying:
You don't think so, you are just logical. "
The physics of altered human consciousness deals with certain conceptualizations that are not easy to enter or view exclusively in the context of ordinary left brain linear thinking. So for borrow the mode of expression of Dr Bohr, some parts of this article will not only require the logical, but intuitive insight into the right brain for get a full understanding of the concepts
involved. Nevertheless, once this is done, I am convinced that their construction and application stand the test of rational criticism.
Paradoxically, having done so much not to try to make judgments based on a frame of reference occult or dogmatic in the end I found it necessary to return, at least briefly, to the question of the impact of experience of the gateway on systems of common belief. I did it because even though it was essential to avoid attempting to make an assessment in the context of such systems, I felt that it was necessary, after completing the analysis, to underline that the resulting conclusions make no violence fundamentally aware of the systems of western or eastern beliefs. Unless this point is clearly established, there is a danger that some people reject the concept of the experience of gateway into the mistaken belief that it contradicts and is therefore foreign to all that they consider to be right and true. This study is certainly not designed to be the last word on the subject but I hope the validity of its basic structure and fundamental concepts in make a useful guide for other staff which are necessary to follow the " bridging " training.

Introduction

In order to describe the Monroe Institute's technique for realization of altered states of consciousness (the experience of the gateway) involving synchronization hemispherical of the

brain, the most effective way to to begin is to briefly describe the mechanics of basis of how the methods work related such as hypnosis, meditation transcendental and biofeedback. It is easier to effectively describe what the gateway is beginning with a brief description of it,
associated techniques that share certain aspects common with the experience of the bridge but which are nevertheless different. This way we can develop an initial frame of reference that will provide useful concepts to explain and understand Gateway by comparison, as and when as we proceed.

Hypnosis

According to the theories of psychologist Ronald Stone and Itshak Bentov's biomedical engineering models, hypnosis is essentially a technique that allows the acquisition of direct access to the sensory motor cortex and to pleasure, centers and brain (emotional) parts lower (and associated pleasure centers) on the right side of the human brain after the successful disengagement of the hemisphere stimulus screening function left brain. The left hemisphere of the brain is auto-cognitive, verbal and linear reasoning component of the mind. It performs the filtering function incoming stimuli by classifying, evaluating and assign a meaning before allowing the passage to
the right hemisphere of the mind.
The right hemisphere, which functions as a component non-critical, holistic, non-verbal and seemingly oriented accept what the left hemisphere passes to him without disputes. Therefore, if the left hemisphere can being distracted out of boredom or reducing to a state soporific, semi-sleepy, external stimuli to include hypnotic suggestions are allowed to pass in the right hemisphere where they are accepted and processed directly. The result may involve a reaction emotional from the lower brain region, sensory / motor, responses requiring involvement cortex, etc. Both the sensory and the motor cortex of the right cerebral part of the brain contain a sequence of points

known as "homunculus" which corresponds to the points in the body.
Stimulation of the corresponding area on the cortex elicits an intermediate response in the game partner of the
body. Therefore, the induction of suggestion that the left leg is numb, if it reaches the right hemisphere uncontested and is returned to the appropriate area of the sensory cortex, will cause electrical reaction that will cause the sensation
numbness. Likewise, the suggestion that the person experiences a general feeling of happiness and well-being would be referred to the pleasure center located in the lower cerebral part or in the cortex of the right hemisphere, thus inducing the feeling of euphoria suggested. Finally, suggestions like the one that informs the hypnotic subject that he appreciates the concentration or the powers of memory would be answered in the right hemisphere by accessing the storage capacity unused information normally retained in the reserves of the selection and control of the left hemisphere. This aspect will become important in the context of the gateway process when attention is given to examining how hypnosis can be used to accelerate progress in beginning of the stages of the Gateway experience.

Transcendental Meditation

In contrast, transcendental meditation works in a markedly different way. In this technical, intense focused on the process of making back up energy in the spinal cord at the end of account what appears to be the creation of waves stationary acoustics in the cerebral ventricles which are then conducted to the gray matter in the cortex on the right side of the brain. As a result, according to Bentov, these waves will stimulate and eventually polarize the cortex in such a way that it will tend to signal the along the homunculus, starting from the toes.
The Bentov biomedical model, as described in a book
by Lee Sannella, MD, titled: Kun dal ini -Psychos is or

Transcendence, states that sound waves are the result of the altered rhythm of heart sounds that are caused by prolonged practice of meditation, and who set up sympathetic vibrations in the walls of the fluid-filled cavities that constitute the third of the lateral ventricles of the brain. In addition, according to Bentov, the states of happiness described by those whose Kundalini symptoms have completed the loop complete along the hemispheres can be explained as a self-stimulation of pleasure in the brain caused by the flow of a current along the sensory cortex. Bentov also notes that most of the symptoms described start on the left side of the body means that it is mainly a development occurring in the right hemisphere.

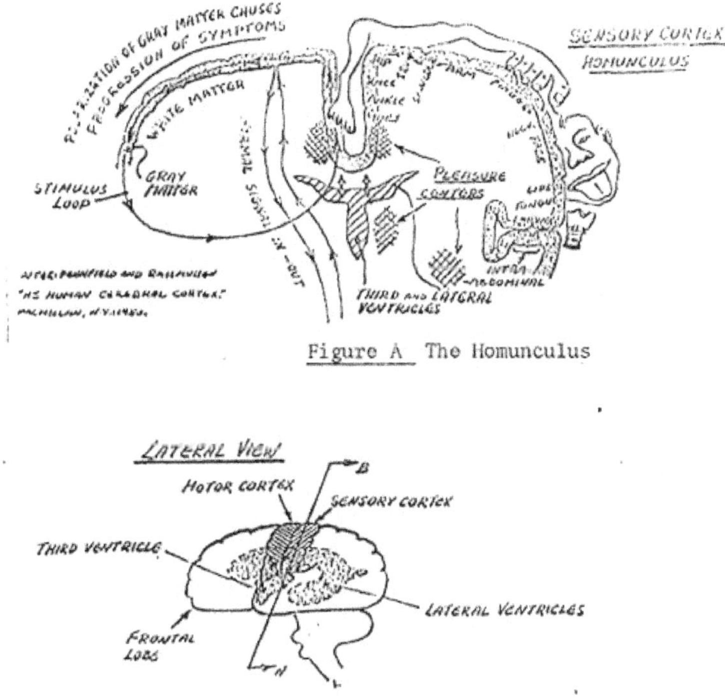

Figure A   The Homunculus

Figure B   The Motor and Sensory Cortex and the Third and Lateral Ventricles

Although normally a period of the practice of meditation for five years or some are required for Kundalini, Bentov states that exposure to mechanical or acoustic vibrations of the range of 4-7 Hertz (cycles per second) for periods prolonged periods can achieve the same effect. Bentov quotes as an example repeated on horseback in a carriage whose suspension and seat combination produces this range of vibrations or is exposed for a long time to periods of time at these frequencies caused, by for example, by air conditioning. He also notes that the cumulative effect of these vibrations may be able to trigger a spontaneous physio-Kundalini sequence in sensitive people who have a particularly sensitive nervous.

Bio-feedback

The third method of modifying consciousness which will be briefly described in biofeedback. The biofeedback is somewhat unique in that it uses in makes the self-cognitive powers of the left hemisphere to access areas of the right brain such as the lower cerebral cortex, motor and sensory and assorted pain or pleasure centers. Instead of delete the left hemisphere as is done in hypnosis, or largely bypassing and ignoring transcendental meditation, biofeedback teaches that the left hemisphere first goes visualize the desired result and then recognize the feelings associated with the experience of successfully accessing the right hemisphere of the specific lower brain, the cortex, pain or pleasure or other areas of the manner necessary to produce the desired result. Of special self-monitoring devices such as the digital thermometer are to be used to inform the left brain when it manages to grasp the hemisphere right in access to the appropriate area. Once done, the left brain can then repeatedly restore the channels involved in order to produce the same measurements external and objective successes. In this way, the ways are reinforced and emphasized to such an extent that the conscience is allowed to access areas appropriate in the right brain using a mode conscious, on demand. For example, if the subject wish to increase circulation in the left leg in order to speed up healing, he can focus with his left brain on achieving this result while carefully monitoring a digital thermometer connected to the left leg. When the concentrated effort begins to succeed, the digital thermometer will register an increase in the temperature of the left leg. At this point, the subject can mentally (left brain) associate the sensations experienced with result obtained and can begin to emphasize, by memory recall, the same process to cause its reinforcement by affirmation and repetition. Of this way, pain can be blocked, healing can be improved, malignant tumors can apparently be suppressed and ultimately destroyed, the centers of pleasure of the body can be stimulated, and a variety of physiological results can be obtained. In addition, biofeedback can be used to speed up greatly the achievement of deep meditative states, in especially for beginners who have no experience in meditative techniques and whose methodology is improved by efficient visualization and external, an objective affirmation. The display of subject's brain waves on a cathode ray tube a proven to be a validated laboratory means by which subjects can quickly learn to place themselves in deeply relaxed states characterized by kind of tranquility and the singularity of the concentration mentality associated with advanced meditation.

Gateway and Hemi-Sync

Now that we have briefly described the basic mechanics of the main techniques for modify or expand the consciousness that some share the objectives and / or methods used in the gateway experience, we can proceed to focus on what this technique involves.
Actually. Basically, the Gateway Experience is a training system designed to bring more strength, focus and consistency the amplitude and the output frequency of the waves brain between the right and left hemispheres in order to modify consciousness, moving it outside of the physical sphere in order to escape the restrictions of time and space. The participant then accesses the

different levels of intuitive knowledge that the universe
offer. What differentiates the bridge experience from forms of
meditation is its use of the Hemi-Sync technic defined in a
monograph by Melissa Jager, trainer of the Monroe Institute, a
state of consciousness defining the EEG patterns of both
hemispheres simultaneously equal in amplitude and frequency. It
also notes that the studies carried out by Elmer and Alyce
Greene at the Menninger Foundation a showed that a subject
with 20 years of training in Zen meditation could establish Hemi-
Sync at will, and the support for more than 15 minutes.
Dr Stuart Twemlow, a psychiatrist and associate of
Monroe Institute research, reports that in our studies of the
Monroe band system on the air cerebral bands, we found that
the bands encourage the focus of brain energy (it can be
measured as with a wattage bulb) in a narrower frequency
band. This focusing of energy is no different from the concept
of yoga that we can translate as one mind.
Dr Twemlow goes on to observe that the individual enters in the
bands beyond the focus, there is a gradual increase in brain
wave size which is a measure of the energy or power of the
brain.

Lamp vs Laser

Melissa Jager uses a metaphor to help clarify the process
involved in using Hemi-Sync in the Gateway experience. She
emphasizes that the human mind in its natural state can be
likened to an ordinary lamp that expends energy in the form of
heat and light but chaotically and incoherent that diffuses its
energy over a large area of fairly limited depth. On the other
hand, the human spirit under the discipline of Hemi-Sync acts as
a laser beam that produces a disciplined flow of light. The energy
flow is projected with total consistency of frequency and
amplitude such that the surface of a laser beam contains billions
of times the energy concentrate found in a similar area on the
sun.

Gateway assumes that once the frequency and amplitude of the human brain are made coherent, it is possible to start accelerating both so that the human spirit soon resonates at the highest levels vibratory. The mind can then embark on the synchronization with multiple energy levels sophisticated and rarefied in the universe. The mind while operating at these increasingly rarefied levels is assumed to be capable of processing the information thus received through the same fundamental matrix by which it makes sense of ordinary physical sensory input to obtain a cognitive sense. Such a meaning is usually visually perceived as symbols but can also be perceived as astonishing flashes of intuition holistic or even in the form of scenarios involving both visual and auditory perception. The mechanism by which the mind performs the function of consciousness will be discussed in more detail later in this article.

Frequency after response. To achieve the synchronization of the cerebral hemispheres, the technic of Hemi-Sync takes advantage of a phenomenon known as frequency After response (FFR), it means that if a subject hears a sound produced at a frequency that emulates one of those associated with the operation of the human brain that will try to imitate the same pattern of frequency by adjusting its brain wave output. Therefore, if the subject is fully awake but hear sound frequencies close to the output brain waves at the Theta level, the brain of the subject will try to modify his wave model brain from normal beta until reaching the Theta level. Since the Theta level is associated with sleep, the affected subject may go from a waking state complete to a state of sleep (provided you do not resist consciously) that the brain strives to train its wave frequency output with the one that the person hears. Since these brain wave frequencies are outside, the spectrum of sounds that can be heard in pure form by the human ear: Hemi-Sync must produce them on the basis of another known phenomenon of the brain's ability to deduce beat frequencies. If the human brain is exposed to a frequency in the left ear which is 10 Hertz below another audible frequency being played in the right ear, rather than hearing one of the two audible frequencies, the brain chooses to hear the difference between them, the beat frequency. So, prevailing of the FFR phenomenon, and using the technic of beat frequencies, the gateway system uses Hemi-Sync and other audio techniques using the FFR phenomenon to introduce a variety of frequencies that are played at an almost subliminal, slightly audible level. The objective is to relax the left hemisphere of the brain, place the physical body in a state of virtual sleep, and bring the left and right hemispheres into the consistency and under conditions designed to promote the production of amplitude and ever higher frequency of wave output cerebral. Subliminal suggestions from Bob Monroe perhaps accompany the different brain waves frequencies, which are sometimes coiled with others sounds to mask the sound frequencies when you wish it. In this way, Gateway strives to provide about the tools allowing him to modify his consciousness based on one's own will over time to through the repetitive use of bands to access, by intuitive means, to new categories information not available to consciousness ordinary.

Role of resonance

However, brain coherence through training in beat the frequencies introduced by stereo headphones are only part of the reason why the gateway system is working. It is also designed to achieve the characteristic physical tranquility of deep transcendent meditative states that result in a complete alteration of the resonance pattern fundamental associated with the frequencies produced by the human body. Zen yoga or transcendental meditation, if practiced long enough, will produce a change in the sound frequency with which the human heart echoes throughout the body. According to Bentov, this resonance change results from the elimination of what the medical profession calls the echo bifurcation for that the sound of the heartbeat can move synchronously and the circulatory system in harmonious resonance about seven times a second.

Bentov describes the role played by the bifurcation echo as follows: When the left ventricle of the heart ejects from the blood, the aorta, being elastic, projects just beyond the valve and causes a pressure pulse to travel along the aorta. When the impulse of pressure reaches the bifurcation in the lower abdomen, a part of the pulse pressure rebounds and begins to move up the aorta. If in the meantime the heart ejects more than blood, and a new pulse of pressure moves down, these two pressure points go possibly collide somewhere along the aorta and produce an interference pattern.

By placing the body in a sinister state, the bands of the "Gateway" achieve the same goal as meditation in that it places the body in a relaxed state that the bifurcation echo slowly disappears like the heart decreases the force and frequency with which it pushes the blood in the aorta. The result is a sound pattern regular and rhythmic sinusoidal that resonates throughout the body and rises in the head in sustained resonance.

The amplitude of this sine wave, when it is measured with a sensitive instrument of type seismograph, is about three times the average of the volume of sound produced by the heart when it is functioning normally.

Brain stimulation

Bentov's biomedical model shows that this resonance is of considerable importance because it is directly transmitted to the brain. The vibration that result is received and transmitted to the brain itself via the third left ventricle filled with fluid located at the above the brainstem. A pulse electromagnetic is then generated which pushes the brain to increase the amplitude and frequency of brain wave output, as observed by Dr.Twemlow in his research on the effects of bands Hemi-Sync. In addition, the brain is contained in a tight membrane which, in turn, is cushioned by a thin layer of fluid located between it and the skull. As coherent

resonance produced by the human heart in a state of deep
relaxation reaches the fluid layer surrounding the brain, it sets up
a rhythmic pattern in which the brain moves up and down
about 0.005 to 0.010 millimeters in a continuous pattern.
The self-reinforcing nature of resonant behavior
explains the body's ability to support this movement
despite the minimum energy level involved, from this
way, the whole body, based on its own micro-
motion, works as a system vibration that transfers energy in a
range of 6.8-7.5 hertz in the ionospheric cavity of the Earth,
which resonates itself at about 7-7.5 Hertz. From this process,
Bentov says: "This happens at a very long time wavelength of
about 40,000 km, roughly the perimeter of the planet. In other
words, the signal from the movement of our bodies travels
around the world to about a seventh of a second across the field
electrostatic in which we are integrated. The wavelength knows
no obstacles.
Naturally, she will pass just about anything: the metal, concrete,
water and the fields that make up our body. It is the ideal way to
transmit a signal telepathic. "
Therefore, the gateway process is designed to quickly induce a
state of deep calm in the nervous system and to lower
significantly blood pressure, causing the circulatory system,
skeleton and all others organ systems and start to vibrate so
consistent at about 7-7.5 cycles per second. The resonance sets
up a regular sound wave and repetitive which propagates in
consonance with the field
electrostatics of the Earth.

Energy training

As the body is transformed into an oscillator coherent vibrating in
harmony with the electrostatic environment, the exercises
included in the gangway bands invite the participant to
increase the energy field surrounding your body, presumably
using the energy field of the Earth that the body is now training in

because of its ability to resonate with it. This puts the
energy field of the body in homogeneity with its environment and
promotes movement of the seat of the consciousness in the
environment partly in response to the fact that the two
electromagnetic medians are maintaining a unique energy
continuum. So the same process that shifts the brain into
concentrate consistency at frequency and amplitude levels
regularly higher in order to train frequencies analogues in the
universe for data collection also promote improved energy levels
body to a point sufficient to allow the subject a movement out of
the body when it is ready to do so. In furthermore, by resonating
with the electromagnetic sphere of Earth, the human body
creates a carrier wave surprisingly powerful in aiding the activity
of communication with other human minds.

Consciousness and energy

Before our explanation can continue, it is essential to define the
mechanism by which the mind human performs the function
known as consciousness, and describe how this consciousness
operates to deduce the meaning of the stimuli it receives. For
to do this, we will first consider the character fundamental to the
material world in which we live our physical existence in order to
perceive with precision the raw materials with which our
conscience must work. The first point to emphasize is that the
two terms, matter and energy tend to be misleading if taken to
indicate two different states of existence in the physical world
that we know. Indeed, if the term material designates
the solid substance as opposed to the energy which is
understood as any force, the use of first is entirely
misleading. Science knows now that the two electrons rotating in
the energy field located around the nucleus of the atom and the
core itself are made up of nothing more than oscillating energy
grids.

Solid matter, in the strict sense of the term simply does not exist

The atomic structure is rather composed of oscillating energy grids surrounded by other oscillating energy grids that orbit at extraordinarily high speeds. In his book, *Stalking the Wild Pendulum,* Itzhak Bentov gives the following digits. The energy network that makes up the atomic nucleus vibrates at about 10 22 Hertz (which means 10 followed by 22 zeros). At 70 degrees Fahrenheit, an atom oscillates at the rate of 10 15 Hertz. A molecule integer, composed of a number of atoms linked between them in a single energy field vibrates in the 10 Hertz range. A living human cell vibrates at about 10 hertz. The human being, the brain, the consciousness are, like the universe, nothing more or less than one extraordinary energy field system complex. The so-called states of matter are in reality of variations in the state of energy, and the human consciousness is a function of the interaction of energy in two opposing fields.

Holograms

Energy creates, stores and recovers in the universe by projecting at certain frequencies in a mode three-dimensional which creates a living model called hologram. The concept of the hologram can be more easy to understand using an example cited by Bentov in which he asks the reader to visualize a bowl filled with water in which three pebbles are fallen down. Like the ripples created by the entrance simultaneous of the three pebbles radiate outwards towards the edge of the bowl, Bentov further asks the reader to visualize that the surface of the water is suddenly jelly, so the ripple pattern is instantaneous. The ice is removed leaving the three pebbles still in place at the bottom of the bowl. The ice is then exposed to a powerful and coherent source of light, like a laser. The result will be a model in three dimensions or representation of the position of the three pebbles suspended in the air. Holograms are able to encode as many details as possible take a holographic projection of a glass of water from swamp and see small organisms not visible to the eye naked when the glass of water itself is examined. The global concept of holography, despite its implications scientists, has only been known to the physicist since the underlying mathematical principles and were developed by Dennis Gabor in 1947 (he later won an award Nobel for his work). Laboratory demonstration of Gabor's work did not take place until years later invention of the laser. As biologist Lyall explains Watson: "The purest type of light that we have is the one produced by a laser, which sends a beam in which all the waves are of a frequency, like those made by an ideal roller in a perfect pond. When two laser beams hit each other, they produce an interference pattern of light waves and dark that can be recorded on a plate photographic. And if one of the beams, instead of coming directly from the laser, is reflected first from an object such than a human face, the result The pattern will be very complex, but it can still be saved.

The recording will be a hologram of the face. "

The part encodes the whole

More importantly is the fact that even though we have dropped our frozen hologram of the model of ripple on the ground and broke it in a number of pieces that each piece would recreate the whole holographic picture all by itself. The smaller the room, the more blur and distortion would be a holographic projection result but the fact remains that an entire projection would nevertheless be made. The key to creating a hologram is that the energy in motion must interact with energy in a state of rest (non-movement). In the previous example, pebbles represent energy in motion while the water (before its agitation by the pebbles) represents energy at rest. For activate or, indeed, to "perceive" the meaning of a holograph, energy (in this case, a source of coherent light like a laser beam) must go through the interference pattern generated by interaction between energy in motion and energy at rest. At simple example given by Bentov, this requirement was filled by maintaining the gel interference pattern in front coherent light to project the three images dimensional holographic (its "meaning") in the space. Like Marilyn Ferguson, editor of Brain / Mind Bulletin tells us: "Another characteristic of a hologram is its effectiveness. Billions of bits of the information can be stored in a tiny space.

The matrix of consciousness

The universe is composed of energy in interaction of fields, some at rest and others in motion. It is in itself a gigantic hologram of an incredible complexity. According to Karl's theories Pribram, a neuroscientist at Stanford University and David Bohm, physicist at the University of London, the human mind is also a hologram that matches the universal hologram by means of exchange of energy thus deducing the meaning and the realization the state we call consciousness. As energy flows through various aspects of the universal hologram and is perceived by the electrostatic fields which make up the human mind, holographic images transported are projected onto these fields electrostatics of the mind and are perceived or understood insofar as the electrostatic field operates at a frequency and an amplitude that can harmonize with and therefore "read" the energy carrier waveform that crosses it. Changes in frequency and the amplitude of the electrostatic field which includes the human mind determines the configuration and therefore the character of the holographic energy that the mind projects to intercept transmissions from the holograpgic universe. Then to understand what the holographic picture means "to say", the mind proceeds to compare the image received with itself. Specifically, it does so by comparing the received image with that part of its own hologram which constitutes memory. By recording the differences in geometric shape and frequency energetic, consciousness perceives. As the saying goes psychologist Keith Floyd: "Contrary to what everything everyone knows, it may not be the brain that produces consciousness - but rather the consciousness that creates the appearance of the brain. "

Brain in phase

The process of consciousness is most easily conceivable if we image the holographic entry with a three-dimensional grid system superimposed on it so that all energy models contained can be described in terms of geometry three-dimensional using mathematics to reduce the data to two dimensional shapes. Bentov says scientists suspect that the human mind works on a simple binary system "Go / no go" like all digital computers. Therefore, once a matrix is superimposed three-dimensional on holographic information that he wishes to interpret and reduces this information mathematically to a two-dimensional form, the mind can treat it completely using its system fundamental binary like any computer made by human hand can process volumes of data and make various comparisons between data and the information stored in its memory. Our minds function in the same way, perceiving by comparison only. Bentov sets out the proposition of this way: "All of our reality is built to make such comparisons whenever we perceive something, we always perceive the differences. »In states of expanded consciousness, the right hemisphere of the human brain in its mode of holistic, non-linear and non-verbal functioning acts as a primary matrix or receptor of this holographic input, operating in phase or consistency with the right brain, left hemisphere provides the secondary matrix, a method of computer-like operation to filter more data by comparison and reduce it to a discreet, two-dimensional shape.

Evaluation

As long as the gateway succeeds in creating a refinement in the energy matrix of the mind, it succeeds in expanding or modifying consciousness so that it can perceive without resorting to the intercession of the senses physical such as the universal hologram can ultimately to be perceived and understood. Marilyn Ferguson wrote that the theories of Pribram and Bohm seem to count for any transcendental experience, the events paranormal and even normal, perceptual oddities ...

She continues to say of Pribram: "Currently, he offers an astonishing and complete model arousing great emotion among those who are intrigued by the mysteries of human consciousness. His holographic model combines brain research with theoretical physics; this explains the normal perception and simultaneously takes the paranormal experiences and transcendental of the supernatural by explaining them in the nature frame. Like some discoveries oddities of quantum physics, the reorientation radical theory suddenly makes sense of the paradoxical sayings of mystics through the ages. "

Self-cognition

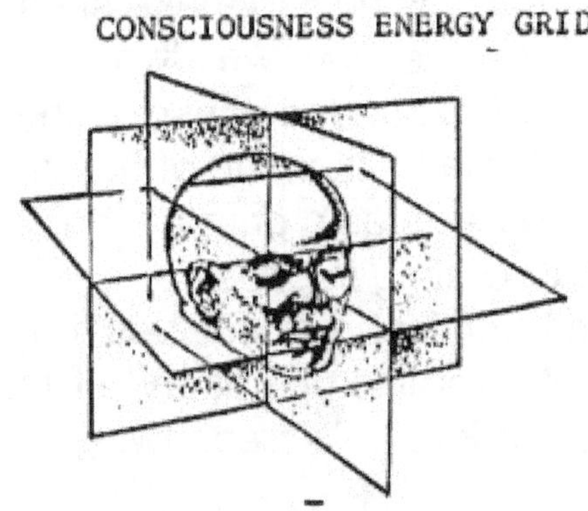

To fill in our outline of the process by which the mind realizes and exercises consciousness, we must also describe the mechanism that counts for the aspect of human thought which differentiates

it from consciousness plants or animals, i.e. knowledge
self. Humans not only know, but they know they know. They are
able to watch themselves even the process of thinking and
maintaining a consciousness of it.
The left hemisphere consciousness grid works like the spirit of
computer software to reduce right hemisphere entries to symbols
and concepts verbal. The Right Hemisphere Consciousness Grid
reduces the three dimensions, a holographic image, to a thought,
the assessment of consciousness being able to do because
there are objective standards that she has adopted. The human
hologram projects an impulse to produce verbal self-knowledge.

Space time dimension

To explain how to access in the dimension the limits of space-
time and understand why the human conscience can do this, we
must first appreciate what the next task should be. Make them
constitute the time and space are necessary for that energy or
force penetrates a transcended being. The physicists defend
that. It is conscious force, without limit. He has no beginning, no
form – one infinite format, the fundamental power.

Intermediate dimensions

In recent years, include infinity (i.e. our boundaries), x physical
existence but we cannot perceive a spatio-temporal dimension in
which we have gradients or dimensions. It is superimposed on
everything like at home through which the energies to enter
these intermediate dimensions, in this state of infinity, (the
Absolute).
It is an aspect of quantum mechanics that applies to the fact that
the entire oscillation frequency (such as a brain wave) reaches
two resting points complete which constitute the limits of each
oscillation individual (i.e. downward). Without these points

rest, an oscillating waveform would be impossible since the rest points are necessary for allow the energy to change direction and therefore continue to vibrate between rigid limits. But it's as true as when, for an infinitely brief moment, this energy. reaches one of its two resting points it clicks space-time and joins infinity. This step critical out of space-time occurs when the oscillation speed falls below 10 -33 centimeters per second (Planck distance). Use the Bentov's words: "quantum mechanics tells us that when the distances are less than the distance of Planck, which is 10 "33 cm, we are indeed entering a new world. »The wave pattern of the human consciousness reaches such a frequency level that the pattern of "clickouts" is so close that there is continuity virtual in it. Then part of that consciousness is postulated to establish and maintain its function as collection of information in the dimensions between space-time and the absolute. Thus, as the continuous pattern "Clickout" is established in continuous phase at speeds above below the Planck distance but before reaching the state of total rest, consciousness passes through the space-time mirror in the manner of Alice who begins his journey to wonderland.
The Gateway experience, with its Hemi-Sync techniques associated, is apparently designed, if used systematically and patiently, to allow the human consciousness to establish a coherent pattern of perception in these dimensions where apply speeds less than the Planck distance. That is true regardless of whether the individual exercises his consciousness while in his or her physical body done after separating this consciousness from the body physical.

Subatomic particles

The behavior of subatomic particles provides a interesting example of the "click out" phenomenon discussed in the previous paragraphs. In an article prepared for Science Digest magazine, Dr John Gliedman mentions how the particles subatomic cells communicate their fields energy are entrained as a result of a collision between them. The communication concerned is, of course, postulated during the "click" in phase "in the oscillation of energy fields including the subatomic of particles involved. It is this cause that explains the cross communication in terms of space velocities temporal, seems to imply excessive speeds light. In reality, the theory of relativity of Einstein is not invalidated but, the communication concerned rather takes place outside the dimension of space-time to which the theory of relativity is strictly contained. Specifically, Dr. Gleidman tells us: "Quantum theory postulates a kind of long-range Siamese, a twin effect whenever two subatomic particles collide and go then to two different routes. Even when the particles are halfway through the universe, they meet instantly to everyone's actions. And in doing so, they violate the prohibition of relativity faster than the speed of light. "

Dimensions Between

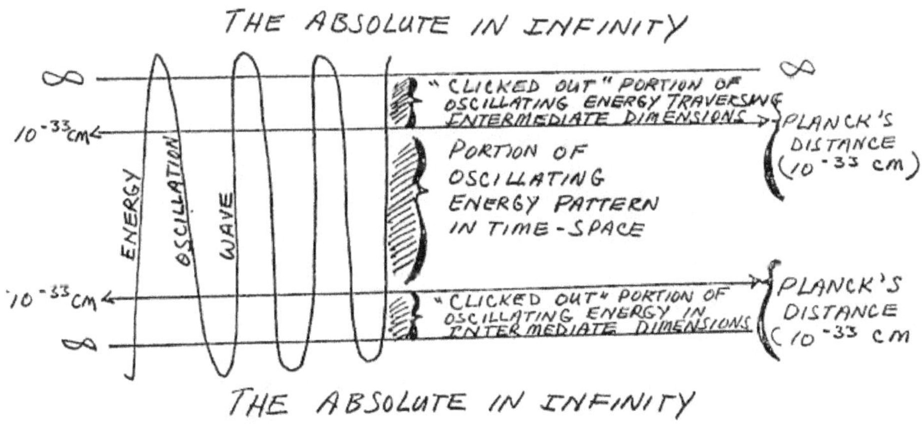

Now that we have postulated the legitimacy of the assertion that the forms of energy that make up the consciousness can go beyond the spatial dimension temporal, we must focus on the forms of energy that inhabit these dimensions between space-time and the Absolute. In doing so, we can better perceive the form that reality takes when it is encounter in intermediate dimensions. In this context, Bentov tells us that: "The relation of cause to effect between the events is broken, the movements become jerky rather than smooth. Time and the space can become grainy or chunky. Perhaps a piece of space can be crossed by a particle material in any direction without necessarily be synchronized with a piece of time. In short, a pair of events will occur in time or space, the pair not being connected causally but by a random fluctuation. "

This means that in the dimension of space-time where the two concepts are applied in such a way generally uniform, there is a proportional relationship between them. Some space may be covered by the energy moving in a particle or a wave train in a certain time assuming a speed specific virtually anywhere in the universe spatio-temporal The relationship is clear and predictable.

However, in the intermediate dimensions beyond of space-time, the limits imposed on the energy for put in a state of oscillating motion is not uniform as they are in our physical universe.

A myriad of various distortions and incongruities are therefore likely to be encountered so that our nice clear assumptions about the relationship between the time and space as we know them in this dimension, do not apply. But even more importantly, access is open to the past and the future when the dimension of the current space-time is left out.

Special status, out-of-body experience

Although human consciousness can, with enough practice, go beyond the dimension of space-time and interface with other energy systems in other dimensions, the whole process is significantly improved if this awareness can be detached to a large extent from the physical body before this interface is attempted. Once an individual becomes proficient in the technique of

movement out of the body and then reaches the point where it comes out of space-time out of his body, he wins the benefit of slamming part of his conscience improved from a base located much more close to the dimensions with which he wishes communicate. In other words, since it starts from a point much higher, to use an analogy in the space-time context, the part of his consciousness involved in clickouts will take much longer to interact in dimensions beyond space time because it takes less time to go through the middle layers. In addition, once the individual is able to project consciousness beyond space time, consciousness would logically tend to drive its frequency output with the new energy environment to which it is exposed, thus greatly improving the consciousness alteration of the individual can be further modified to reach a much higher point of focus and a very refined oscillating pattern.
Consequently, a self-reinforcing process should ensue, in which consciousness in the state outside body can be projected beyond the dimension of space-time, moreover its level of production energy would be improved, thus promoting travel potentials. The provisional conclusion to be made is that the outside of the body can be seen as a means extremely effective in speeding up the process awareness enhancement and interfacing with dimensions beyond space-time. If the practitioner of the Gateway technique has the choice to focus on achieve and use the experience outside the body rather than to concentrate his efforts to develop his conscience exclusively from a physical basis, the first seems to promise much faster success and impressive than second choice.

The Absolute in perspective

It may be helpful at this point to take a break and recap the main aspects of our intellectual trip of space-time in the realm of the Absolute.
We talked at length about the hologram incredibly complex that is created by the intersection energy models generated by the

totality of all the dimensions of the universe, space-time included. We noted that our minds constitute fields energy that interact with various aspects of this hologram to deduce information that is eventually processed in the left hemisphere of our brain to reduce it to a form that we use for the process we call thinking. We hinted that this hologram is the finite incarnation in the form of active energy of the infinite consciousness of the Absolute. This is the title that we have assigned to this vast reserve of energy in a state of perfect rest in which the physical universe is in layers, and where it came from. Incidentally, to describe this, Bentov uses the analogy of a very deep sea, comparing the depths of the sea to the dimension of the Absolute by assigning the waves launched by the storm to represent the physical universe with which we are familiar. The slightly choppy currents of the sea lies between the turbulent surface and the totally still depths represent energy in the process of resting or coming out of rest.

From Big Bang to the torus

From the widely accepted theory of the Big Bang, Bentov presents a conceptual model to describe the process of evolution of space-time position relative of the universal hologram. This hologram is often called torus because it is believed to have the general shape of a huge, self-contained spiral. Basing its thesis on recent studies concerning the distribution quasars (quasi-stellar objects), and operating on the principle that in the universe, the smaller processes have tend to be mirror images electron model around the nucleus of an atom reflects how the planets orbit their suns, etc.
Bentov postulates the following scenario. Taking his cock the observed capacity of quasars to eject enormously concentrated bundles of matter from their interiors in a controlled and non-concentric version of the Big Bang, he envisions a similar process producing in the generation of the universe. Noting that

those galaxies located north of our own galaxy move away faster
than those to the south, and that those to the east and west are
clearly more distant, Bentov considers that as proof substantial
than the stream of matter which has extended in our universe
has turned on itself, forming possibly an ovoid or egg shape.
He sees "matter" in our universe entering the ovoid model after
ejection of a compound nucleus of extremely compressed energy
through a white hole. At the end of its journey to the end of the
ovoid, it sees him leave through a black hole. In such a model,
we observes that time is a measure of change which
occurs as energy evolves into new more complex shapes as they
are progression on the side of the white hole of the nucleus,
around the shell of this cosmic egg until it penetrates in the black
hole.
In other words, as energy expelled from infinity and confined
within limits by the consciousness of the Absolute achieves
shape and movement after ejecting the hole white on top of the
egg, time starts like a measure of the cadence of this
evolutionary movement in as "reality", an eggshell on its way to
the black hole.

Our place in time

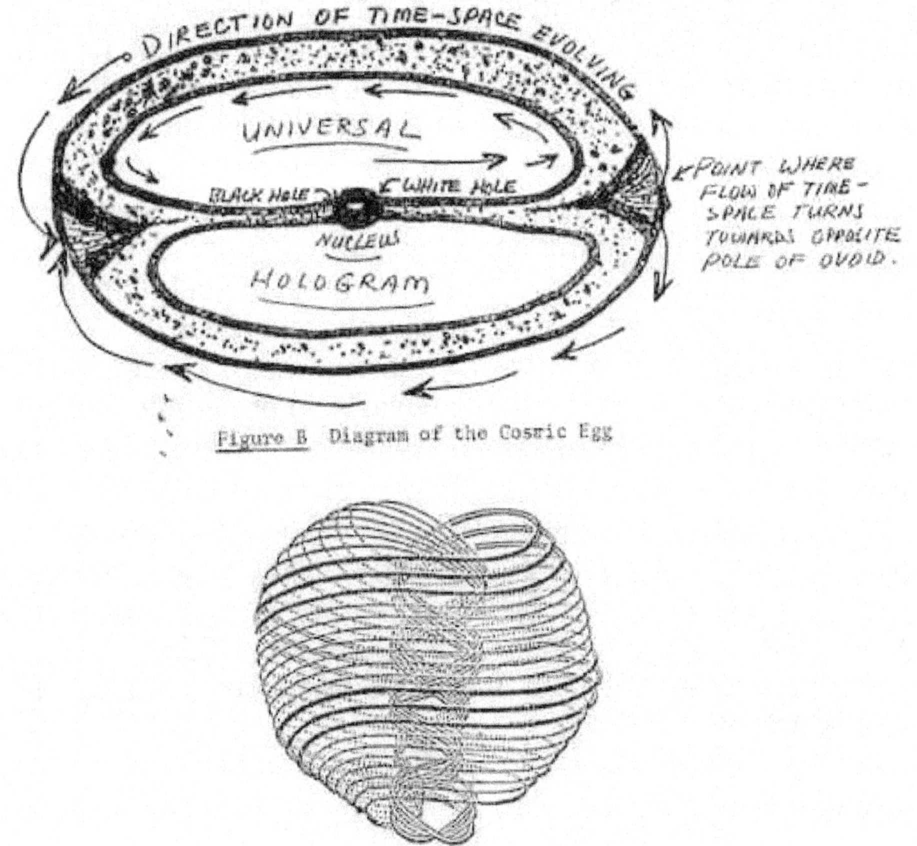

Figure B Diagram of the Cosmic Egg

Figure C Stylized Rendition of a Simple Torus

The observed distribution of galaxies suggests that our particular universe is located near the top of the egg at the point where matter begins to fold, explaining thus the reason why the galaxies where we see that the north is moving faster due to the flow of matter towards the far end of the cosmic egg. At-above this egg is the Absolute which sustains the radiation, nucleus from which the jet of matter original comes out.

As the flow of matter moves around the ovoid to its destination to the black hole where it will be reabsorbed in the radiant nucleus and then the Absolute, it generates the interference pattern in the cosmic egg which constitutes the universal hologram or torus.

154

Since the torus is generated simultaneously by the matter in all
the different phases of time, it reflects the development of the
universe in a past, the present and future (as we would see in
our particular perspective a time phase). In reflecting on this
model, it becomes possible to see how human consciousness
brought to a state sufficiently modified (focused) could obtain
information about the past, present or future because they all
exist in the universal hologram simultaneously (in the case of the
future because all consequences of the past and present can be
seen come together in the hologram such that the future may
be predicted or seen with complete accuracy).
In addition, it is possible to see how the implosion of energy
patterns intersect and create a hologram in four incredibly
complex dimensions or torus, in the shape of a spiral in reflection
of the evolution model multidimensional developing. All of the
movements of the energies that make up the universe
leave their mark and therefore tell their story over time.

Quality of consciousness

We noted earlier that the state outside the body involves the
projection of a major part of the model of energy that represents
consciousness so that it can move freely throughout the Earth's
sphere for any purpose information acquisition or in other
dimensions outside of space-time, maybe to interact
with other forms of consciousness within the universe.
Consciousness is the principle of organization and support which
provides the impetus and direction to bring and keep the energy
moving in a given set parameters so that a specific reality
results.
When consciousness reaches a state of sophistication
in which he can perceive himself (his own hologram)
reaches the point of self-knowledge. The beings humans have
this form of consciousness as the Absolute, but in the second
case, it is a function of energy and its quality of consciousness
associated with infinity (omniscience and omnipotence in unity

perceptual). When the energy returns to a state of rest
total in the Absolute, it returns to the continuum of consciousness
in the pool of unlimited perception and timeless that resides
there. Thus, no longer a system energy in the material state, the
more consciousness to maintain its reality. Our conscience is this
differentiated aspect of consciousness universal which resides in
the Absolute. It counts for the organization of the energy patterns
that constitute our physical body but is distinctly distinct and
higher than her. Since consciousness exists outside of reality,
beyond the limits of space-time, it, like the Absolute, has neither
beginning nor end. Reality has both a beginning and an end
because it is limited in space time, but the fundamental quantum
of energy and its associated consciousness is eternal. When
reality is ends, its constitutive energy simply returns to
infinity in the Absolute.

Consciousness in perspective

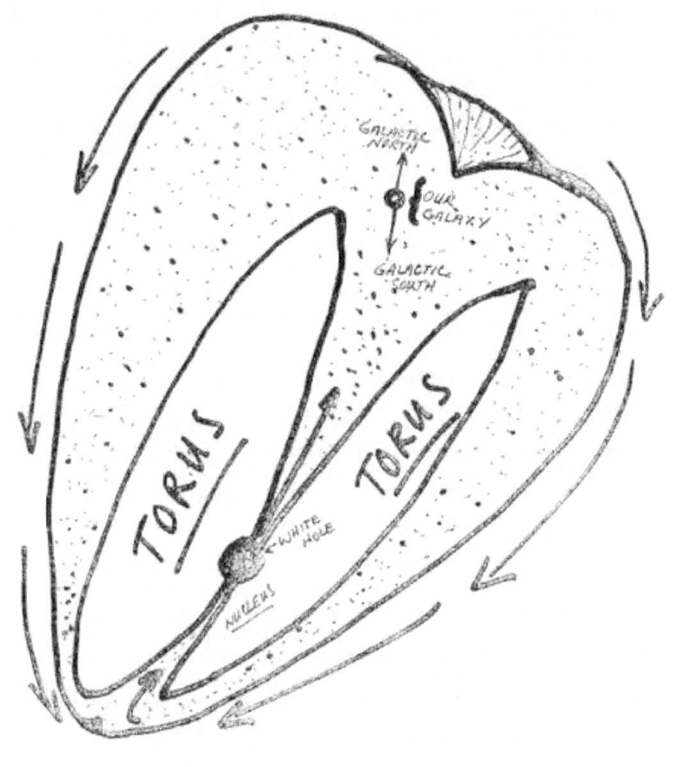

Having found that human consciousness is capable to separate from physical reality and interact with other intelligence from other dimensions in the universe, and that it is both eternal and ultimately destined to return to the Absolute we are faced with the question: So, what is happening?

Since memory is a function of consciousness and therefore enjoys the same eternal character as the consciousness which explains its existence, it must be admitted that when consciousness returns to the Absolute, it brings with it all the memories accumulated thanks to experience in reality. The return of consciousness to the Absolute does not imply an extinction of the separate entity that consciousness organized and sustained in the reality. It suggests a differentiated consciousness which merges and participates in universal consciousness and infinity of the Absolute, without losing the distinct identity and accumulated self-knowledge that his memories confer. It loses the ability to generate independent thought holograms, since this can only be done by energy in motion. In other words, it retains the power to perceive but loses the power of will or choice. In exchange, however, this consciousness participates in the infinite continuum of omniscient consciousness which is a characteristic of the energy of the present. Therefore, when a person experiences a state outside the body, the spark of consciousness and memory which is the ultimate source of his identity is projected to let him play and learn dimensions both inside and out outside of the space-time world in which his physical component has a short period of reality.

Gateway method

Having put the experience of the gateway into context by postulating a structure of how, why it seems function, and shown how to achieve it, the time is come to examine the specific techniques that understands the bridge training process (Gateway). These techniques are designed to allow the user of the Gateway bands to manipulate the states high energy levels that can be achieved if the user continues to work with the bands over a period of time. The time needed to reach states energy sources and make full use of techniques vary with the individual. The sensitivity of his nervous system, its general state of mind and the extent in which he may have previously developed ease in related techniques such as transcendental meditation are all factors affecting how quickly it can be expected to progress. The process begins by learning to each participant to isolate the concerns in a viewing device called a conversion box of energy. Then the participant is introduced to a method to encourage his mind and body to achieve a state of resonance by pronouncing a single tone, a monotonous, prolonged buzz that creates a sensation of vibration especially in the head. He engages in this "resonance chord" as we calls it humming with a chorus of such sounds which are contained on the Gateway band. Following this, the participant is exposed to the assertion of the gateway and is encouraged to repeat it to himself as he sees fit repeated on the tape. This statement is a statement to the effect that the individual realizes that he is more than a simple physical body and which he deeply desires to develop his conscience.

Hemi-Sync introduced

After that, he is exposed for the first time to Hemi-Sync sound frequencies are encouraged to focus and develop a perception and appreciation of those feelings that accompany synchronization of the resulting brain waves.
Then comes the technique of progressive and systematic physical relaxation while the Hemi-Sync frequencies are extended to include additional shapes of pink and white noise designed to put the body on the virtual sleep threshold as well as to calm the left hemisphere of the mind while raising
the right hemisphere in a state of heightened attention. A
once all this is done, the participant is invited to consider the creation of an energy balloon composed of an energy flow starting at the center of the top of the head and descending in all directions to the feet. The energy involved in this flow then passes to through the body and spring into the model of energy balloon, which sets up a pattern reminiscent of much the cosmic egg discussed earlier. No only it improves the flow of body energy and encourages early achievement of a resonant state suitable, but it is also designed to provide protection against conscious entities possessing lower energy levels than the participant can meet in the event that it achieves a state out of the body.
It serves a precautionary purpose in the unlikely event that the first experience outside the body involves a direct projection outside the sphere of the Earth.

Advanced techniques

Having reached focus 10, the participant is now ready to strive to achieve a state of consciousness broad enough to actually start interacting with dimensions beyond those associated with its physical experience reality. This state is called Focus 12 and involves conscious efforts on his part while additional forms of "pink and white noise" enter the sound stream being directed into his ears of the Gateway band. Once the participant has reached this greatly expanded state of consciousness, it is ready to start using a series of techniques specific or "tools" as characterized by the Monroe Institute allow him to manipulate his expanded consciousness newly found to get practice, back useful information to promote the discovery of self and personal growth.

A – Problem solving

This technique consists of identifying the problems fundamentals that the individual wishes to see resolved, filling his expanded consciousness with his perception of these problems and then project them into the universe. In this way, the individual seeks help from what the Monroe Institute its hiqher self, in other words its expanded consciousness, to interact with the hologram universal to get the information you need to problem solving approach can be used to solve personal difficulties, problems technical field of physics, mathematics, practical administrative problems, etc.
The answers to the problem-solving technique can be received almost immediately, but often they are based on the development of intuition over the next two or three days.
Often the answer comes in the form of a sudden and holistic perception the individual finds suddenly he just knows the answer in all its ramifications and completely in context, sometimes even without being able to put his perception

newly found in words, at least initially. In some cases the answer may even come in the form visual symbols that the individual will see with his mind while in Focus 12 state and will need to interpret after returning to normal consciousness.

B – Modeling

This technique involves the use of consciousness to achieve the desired goals in the physical sphere, emotional or intellectual. It focuses the concentration on the desired goal in a Focus state 12, extension of individual perception of this goal in the expanded conscious set, and its projection in universe with the intention that the desired goal is already a successful business that is intended to be achieved in the specified time frame. This particular methodology is based on the belief that the thought patterns generated by our consciousness in a state of expanded awareness create holograms that represent the situation that we wish to bring and in doing so establish the basis for the effective achievement of this objective. Once the hologram generated by the thought of the lens sought after is established in the universe, it becomes an aspect of reality that interacts with the mind the hologram universal to achieve the desired goal that may not not, under other circumstances, occur. In other words, the structuring technique recognizes the fact that since consciousness is the source of all reality, thoughts have the power to influence the development of reality in space-time applies to us if these thoughts can be projected with adequate intensity. However, the higher the objective sought is complicated and the more it deviates radically our current reality, the more the universal hologram will have need to reorient the sphere of reality to accommodate our desires. Monroe Institute trainers highlight guard against trying to force the pace of this process because the person might be successful by moving their existing reality with drastic consequences.

C – Color breathing

The next technique is called *breathing color* and is designed to use the awareness enlarged and highly focused attention associated with the Focus 12, the state of imagining different colors particularly intense and lively in order to use them for resonate with and in turn to activate the body of its own energies. Basically, in terms of practical application, it is a healing technique which is designed to restore the body and to improve its physical capacities by balancing, revitalizing and re-tuning the bodily energy flows. It is based on the principle that the body's electromagnetic field is able to modify its resonance model in order to drive the energy of the Earth's electrostatic field for its own use. The different colors envisioned in the imagination as part of the technique prompts the mind to know which frequencies and what specific amplitudes are desired in relation to this training and the modifications subsequent in bodily energy flow patterns.
This color has the ability to affect the human mind and is well known for its healing effectiveness. Through example, the application of intense blue light on an area of physical tumescence leads to a rapid observable reduction in swelling while red and, to a lesser extent, yellow makes all the effect reverse. However, in the Hemi-Sync app, the technique of external light sources are not involved, only the mind is the agent of healing and revitalization.

D – Energy bar tool

Magic wands are part of folklore and occult practices of many cultures. The scepters, the staff and the masses carried by the monarchs and high priests perform with such frequency in the history of past eras to suggest that at the very least these elements are type aspects of archetypal symbol that the human mind seems appreciate, perhaps quite subliminally. In anyway, the technique of the energy bar tool involves considering a small point of light intensely pulsed that the participant loads into his imagination with tremendous energy until it is virtually pulsed. The participant extrudes the point under the shape of a vibrating cylinder and sparkling with energy that then uses to channel the force of the universe into selected parts of their body for healing and revitalization.

E – Remote viewing

In addition, the energy bar tool is used as a portal to initiate a tracking technique called "visualization remotely ". In this context, the participant turns his energy bar in a whirlpool through which it sends his imagination in search of new ideas illuminating. The apparent purpose of the symbolism involved in the vortex seems to be spotting the subconscious and give him instructions as to what the participant wishes to do but in terms of symbols nonverbal that the right hemisphere of the mind is capable of understanding.

F – Map of the living body

This technique provides amplification for application of energy bar tool as a way heal specific areas or systems of man body. The configuration of the participant's body is imagined then the different of the major systems such as the nervous and circulatory systems are considered in appropriate colors within the limits of outline held in the imagination. The bar tool energy is then applied to energy, balancing and the healing of the way the participant longed for. In the process, the participant visualizes various flow of colored energy exiting the tool in the system organ or area on which the application revitalizing or healing is done. Since the colors are the result of different wavelengths light, that is to say energy at different frequencies, this technique works on the assumption since the human body is

made up of energy, it can be vitalized and healed by the additive
application of energy additional provided energy is applied
in the appropriate form.

G – Focus 15 – Journey into the past

All the above techniques are carried out at the level of expanded
awareness known as Focus 12. However, the technique of time
travel involves further expansion of consciousness thanks to the
inclusion of additional sound levels on Hemi-Sync bands. Some
sounds are an intensification of the Hemi-Sync base, the
frequencies being designed to further modify the frequency and
the amplitude of the brain waves.
Other aspects of the added sound patterns seem be designed to
provide suggestions almost subliminal to the mind as to what is
desired by way of of expanded consciousness to support
suggestions verbal and instructions also contained on the
bandaged. Even the instructions are highly symbolic, over time is
visualized as a huge wheel in the universe with different spokes
including each gives access to a different part of the past of the
participant. Focus 15 is a very advanced condition and is
extremely difficult to achieve. Probably less than five percent of
all participants actually realize fully Focus 15 state over the
seven days about workout.
Nonetheless, trainers from the Monroe Institute claim that with
enough practice, Focus 15 can be achieved. They also state that
not only the individual's past history is available for review
by the one who produced Focus 15 but other aspects of the
past with which the individual himself has had no connection
can also accessed.

H – Focus 21 – The future

The latest and most advanced of all Focus associated with
Gateway training program involves a movement outside the
limits of space-time as in Focus 15, but paying attention to the

discovery of the future rather than the past. The individual who has reached this state has reached a truly advanced level. Except in unusual circumstances, it is probably impossible to reach except by those who are conditioned through the application of meditation or by those who have long practiced the use of Hemi-Sync cassettes for a period of several months, see several years.

Movement out of the body

This remarkable phenomenon was saved for discussion in detail until the last because of the interest he causes and special circumstances involved in its realization. Monroe Institute points out that the Gateway program was not established only in the purpose of enabling participants to obtain the status out body and the program does not guarantee that most of participants will succeed in doing so during the training at the institute. Only one band among the many that make up the Gateway experience are devoted to the techniques involved in movement extra-corporeal. Basically these techniques are simply designed to facilitate the completion of condition out of the body when its brain wave pattern and his personal energy levels have reached a point that he is apparently in harmony with his surrounding electromagnetism as he feels he has reached the threshold where separation is a possibility. To facilitate achievement of the out-of-body state, Bob Monroe, the founder of the Monroe Institute, is quoted in an article by recent magazine saying that in order to help the participant, the Hemi-Sync tape uses beta signals of approximately 2877.3 CPS (cycles per second).
Since 30 to 40 CPS are considered to be the normal range of brain wave beta signals (those associated with the waking state), it is obvious that the Monroe Institute is convinced that the same increased condition brain wave frequency which promotes altered states of consciousness is also an important consideration in helping to obtain statements out of the

body. The actual techniques used to to separate from the body
involve maneuvers too simple that roll, lift like a pole
telephone in which the individual separates and slides
through the extremities of his body.

Role of REM sleep

It is interesting to note that Bob Monroe informed the
Gateway class which ended on May 7, 1983, that a former
trainer of its operations in Charlottesville, Virginie, found that he
could guarantee movements outside the body by bringing
participants back in a state of rapid sleep and then use the
Hemi-Sync tape technique. It could be depending on the fact that
most people, otherwise all, are deemed to be in a state outside
the body for REM sleep.
REM sleep is the deepest level possible from ordinary sleep and
involves the releaseof the body's motor cortex runs from the neck
down and the complete suppression of consciousness in
the left hemisphere of the brain. The goal is to put the
body in a state of complete immobility in what concerns the
skeleton of muscle structure is concerned, thus further promoting
the state of rest deep necessary to eliminate the bifurcation
echo. Of more, it leaves the right hemisphere of the brain free of
respond to the instructions and suggestions contained in the
Gateway band. However, the use of bands Hemi-Sync at this
point may be less of a factor in realization of the state outside the
body that it is a matter to concentrate the brain enough so that
the residual memory have naturally reached an out state
of the body brought to the waking state. Indeed, it can even
be postulated that some dreams associated with levels
of deep sleep are in fact a function of the same
kind of altered consciousness involved in the interaction
with the universe which plays a role in all states of
Focus 12, 15 and 21 described above.
The difference between these states and the condition of the
mind in REM sleep seems to be that the hemisphere

left is almost completely free from the latter experience such as remembering what has been achieved in altered states of consciousness, and cannot usually not be picked up by conscious desire because the left hemisphere has no knowledge of its existence or its location in the right hemisphere. Certainly some people can be trained to remember their dream state through intense conditioning in the waking state, but
even this may be more of a function of establishing pathways in the right hemisphere, that the left hemisphere can access the next re-entry in the standby state that it is an indication of any conscious involvement specific of the left hemisphere in the process during REM sleep.
In any event, the three apparent conditions required to voluntarily induce a state outside the body in most individuals appears to be:
1. the achievement of a state of deep silence in the body such that the bifurcation echo fades and resonates at about 7 Hertz is established,
2. the synchronization of the two brains, the shapes hemisphere waveform and 3.the subsequent stimulation of the right hemisphere of the mind to achieve a state of heightened alertness (which, of course, interferes with the synchronization of the cerebral hemisphere but not before a sufficient level of the range of frequencies has been established to help achieve the off state from the body).

Potential for information gathering

The information acquisition potential associated with the condition out of the body seems to attract the most attention of the point for the development of practical applications for the Gateway technique. Unfortunately, although the state outside the body can apparently be reached by many people without spending too much time nor efforts, the purposes may currently be limited by the fact that although individuals in this state may travel anywhere on an instant basis in the

terrestrial or in other spheres, the distortion of information in the first context remains a problem major. concern. To date, according to one of the trainers from the Monroe Institute, many of the experiments have been carried out with people
moving from coast to coast in the out-of-body state to read a series of ten computer-generated numbers in a university laboratory. Although most have acquired enough figures to make it clear that their consciousness was present, no one has ever managed to get the ten correct. this seems to be function of the fact that the physical reality in the present is not the only holographic influence that the individual may encounter in an out-of-body state. There is also energy patterns left by people or events occurring on the same physical plane seen,
but from the past rather than the present. Moreover, since thoughts are the product of energy, patterns energies are a reality, it is also possible that encounter thought forms in a state outside the body that mixes with physics, reality is not
easily differentiated. Finally, as Melissa Jager writes, there is another potential problem in that the holograms can be visualized pseudo-scopically, that is to say upside down or as well as they can be seen from the proper perspective.
Some of the distortions occurring at the end of the
intends to prove this cause because in a state outside the body an individual can perceive the patterns of holographic energy emitted by people or things and interact in the reality of space-time under somewhat distorted shape.

Belief System Considerations

In 1967, Alexandra David-Neel and Llama Yongden have wrote a book called *Secret Oral Teachings in Tibetan Buddhist sects* , whose quotation following is taken:
"The tangible world is movement," say the masters, not a collection of moving objects, but the movement itself. "

This movement is a continuous and infinitely rapid flashes of
energy (in Tibetan "tsal" or "Shoug"). All objects perceptible to
our senses, all phenomena, of whatever nature, which
whatever their appearance, are made up of a succession
fast instantaneous events.
The classic description of the universal hologram is found in a
Hindu sutra which says:
"In Indra's sky, they say there is a network of pearls so arranged
that if you look at one, you see all others reflected. "
I took this quote because it shows that the concept of the
universe, at least some physicists come now to be accepted, is
essentially identical aspects with that known to the learned elite
in certain high-level civilizations and cultures in the ancient
world. The concept of the cosmic egg, by example, is well known
to researchers familiar with the ancient writings of Eastern
religions. Theories presented in this article are not the essential
principles of the Judeo-Christian current of thought. The concept
of visible reality (i.e. the "created" world) as an emanation of a
whole mighty and omniscient divinity who is completely
unknowable in its primary state. The Absolute at rest
in infinity is a concept straight out of philosophy
Hebrew mysticism.
Even the Christian concept of the Trinity shines through the
description of the absolute as presented in this article.
The description of energy totally at rest, to infinity corresponds to
the Christian metaphysical concept of the Father while the infinite
self-consciousness residing in this energy provides the strength
of will to put into movement some of this energy to create the
reality corresponding to the son. It is because for achieve self-
awareness, awareness of the Absolute must project a hologram
of itself and then the perceive. This hologram is a mirror image of
the absolute in the infinite, and always exists outside the
time and space, but is in a step of Absolute and is
the real agent of all creation (all reality).

Motivational aspect

It is a step-by-step procedure that involves a
repetitive practice of the techniques concerned, in using each
new insight as a way to penetrate further during the next practice
session.
But the pace of progress is so much faster with the Gateway
approach than with meditation transcendental or other forms of
self-discipline mental and its horizons seem to be much more
wide than the discipline necessary to practice it seems to be
within the reach of even a skeptic of our society. Unlike yoga and
other forms of oriental mental discipline, Gateway does not
require infinite patience and total personal enslavement nor a
faith in a system of discipline designed to absorb all the energies
of the individual for the most part of his life. Rather, it will start
producing at least minimal results within a relatively short period
of so that sufficient feedback is available to motivate and
energize the individual to continue to work with it. Indeed, the
speed at which an individual can expect progress to appear less
in depending on the number of hours spent practicing that a
matter of how quickly he or she is able to use the knowledge
acquired to free the anxieties and stresses in both the mind and
the body.
These energy blocking points seem to constitute the main
obstacle to achieving energetic states and concentration of the
mind necessary for a rapid progression. The more compulsive,
the more the individual can be stuck at the beginning, the more
obstacles he will have the more he will initially encounter a deep
experience or immediate, but the ideas start to come and the
blockages begin to dissolve, the way forward becomes more and
more clear and the value of Gateway change from the status of
an evaluation question intellectual to personal experience.

Conclusion

There is a rational basis in terms of science physical, parameters to consider the gateway as plausible in terms of its essential purpose. Of intuitive intuitions not only personal but practices and the professional character seems to be in the limits of reasonable expectations. However, a phased approach to enter catwalk experience would seem to be necessary if the time to reach advanced states of the altered consciousness must be brought into greater limits easy to manage in terms of establishing an organization-wide operation of my gateway.

The most promising approach suggested in the study previous concerns the following steps:

A. Start by using the Gateway Hemi- bands Sync to obtain a better concentration of the brain and to induce the synchronization of hemisphere.

B. Then add high frequencies of sleep paradoxical to induce brain quiescence left and deep physical relaxation.

C. Provide a hypnotic suggestion designed to allow an individual to induce a self-hypnotic state deep at will.

D. Use self-hypnotic suggestion to achieve a much larger center of concentration and the motivation to progress quickly through exercises Focus 12.

E. Then repeat steps A and B after using the self-hypnotic suggestion that a movement out of the body to happen and remember.

F. Repeat step E for ease of getting a state outside the body under conscious control. Modify the hypnotic suggestion to emphasize the ability to consciously controlling movements outside the body and maintain them even after the end of the state of paradoxical sleep.

G. Approach to Focus objectives 15 and 21 (escape space-time and interact in new dimensions) from an extra-corporeal point of view.

H. Use a multi-focus approach to solve the problem problem of distortion in travel systems land-based information gathering. This approach involves the use of three people in the out-of-state body, we visualize the target object here, in space-time, viewing it in Focus 15 as it slips into the immediate past, and looking at it in Focus 21 as it slips from the immediate future. Recap the three and compare the data collected from the three points of view. If care is taken to ensure that the three come out of the body together, in the same environment, their consciousness energy systems should resonate in a sympathetic oscillation. They can connect to the same planes (dimensions) with greater efficiency.

I. Encourage the search for self-knowledge by everyone involved in the experiments to improve objectivity in observation outside the body and thinking and eliminating personal energy blockages likely to delay rapid progress.

J. Be intellectually ready to react to possible encounters with intelligent forms of energy and not bodily when the limits of space-time are outdated.

K. Arrange for groups of people in the state of Focus 12 unite their consciousness to build holographic patterns around areas sensitive to repel a possible presence not desired.

L. Encourage more advanced participants in the catwalk to create successful holographic patterns and rapid progress for advanced colleagues to assist them as they progress through the bridge system.

If these experiments are carried out, we must hope that we will really find the gateway and the domain practical application for the whole system of techniques that compose it.

12 / When robots become aware of themselves

QBO

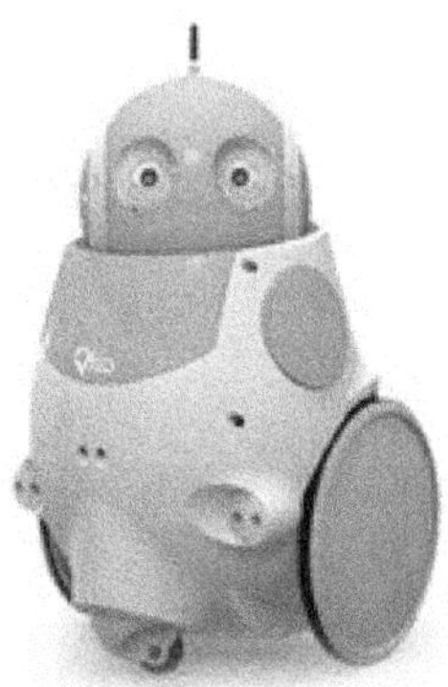

The QBO robot is the first robot to become aware of who he was
during a test. The operator placed it in front of a mirror, and
asked him who he is.
QBO did not know what to answer. Following this, the operator
him pointed out that what he saw in the mirror was his
reflection. QBO then analyzed what it saw and its bank of
data. When the operator asked him the question again, QBO this
time replied that it was he who reflected in the miror. It only took
a few seconds for this robot to understand and analyze what was
in front of him.

NAO

NAO robots, smaller than QBO, also have participated in a little experiment. To test them, an operator informed them that some of them had swallowed a mind numbing pill preventing them from being able to speak. From then on, the operator asked them which pill they had taken. One of the three robots stood up and signaled him that he didn't know which pill he had taken. A few seconds later, the NAO apologized after realizing that he could speak and that therefore his was a placebo. It only took a few seconds to realizing that he himself could speak. A take of new consciousness and without any outside help.

AI-DA

Aidan Meller, Art Gallery Director at Oxford, had the idea of creating Ai-Da eight years ago, naming it in tribute to the English pioneer of science IT, Ada Lovelace. In 2017, the design of the humanoid begins in the workshops of Engineered Arts. With the help of Aidan Meller, engineers created the skeleton, then gave Ai-Da a feminine appearance in order to correct the imbalance between men and women, omnipresent in the art world. The students at Leeds and Oxford universities have developed the robot's brain made up of algorithms complex. Thanks to its programming, Ai-Da can to converse and choose, without the help of man, the works that she wishes to draw. " *We don't know what she's got head when she begins to sketch. It is impossible to predict what it will achieve,* " continues the merchant of art. As Ai-Da only knows how to hold a pencil and realize simple sketches, she provides, to complete her works, indications to human painters who apply.

13 / Distinguish between reality and fiction

For many theorists, such as Thomas Pavel, Marie-Laure Ryan and Françoise Lavocat, Professor at Sorbonne-Nouvelle University, fiction is a matter of degree: there are no borders clear between fiction and "reality".
The philosopher Hans Vaihinger developed in 1911 in The Philosophy of `` *as if* '', a theory according to which all knowledge, even scientific, is only fiction. We know the phenomena, we build scientific models, but we cannot know the essence of things.
Today in 2020, we are closer to realize that we live in a simulation that we do not think and we are hardly able to distinguish the true from the false. Take the example of virtual influencers.

You think this handsome man with pepper hair and salt is real? Nope! He is a virtual influencer created by Roarty Digital for the famous KFC fast food. Isn't that amazing? The content of a virtual influencer would be three times more attractive than that of an influencer real! So a human should do almost four times more Instagram posts to get the same number of followers as avatars.

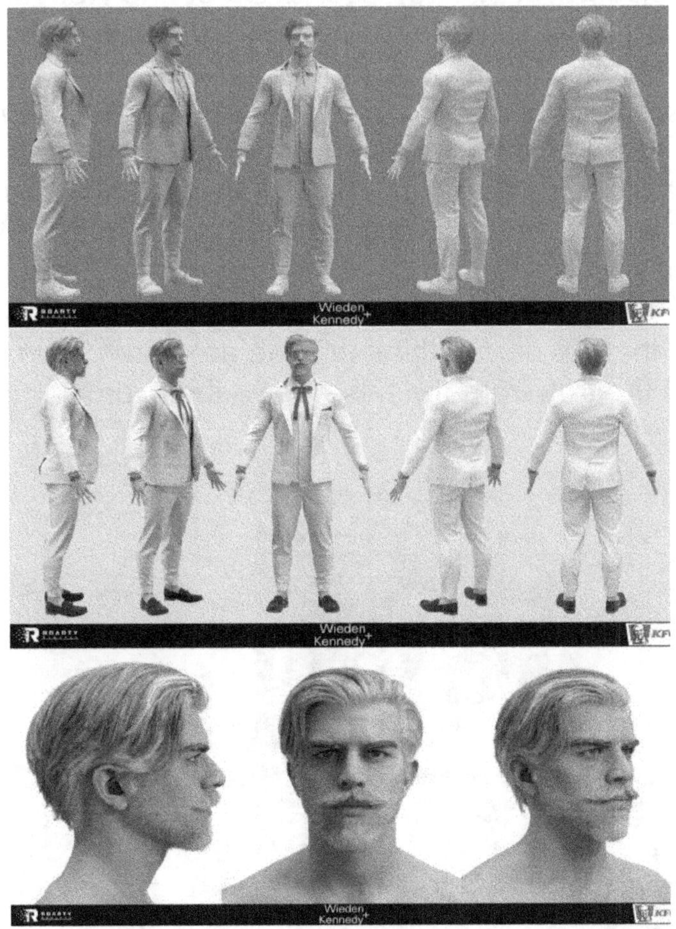

Some big luxury brands are snapping up virtual influencers because in addition to benefiting from high visibility, these fictional characters are more malleable than human influencers. They are totally controllable, this which is very popular with advertisers.
One of the societal concerns is that they could take jobs from real models. So the false weighs heavier than the true! Indeed the market influencers is huge and lucrative; 5 billion dollars

were spent on influencer marketing Instagram in 2018, and the trend is only increasing.

" There isn't really much of a difference between a real person and one of those CGI accounts, " explains Buzz Carter, Marketing Specialist digital at Bulldog Digital Media. *" If the interaction and confidence are there, a campaign with Lil Miquela should be as effective as with any another influencer. It is also easier because the virtual influencers don't need to travel by plane for photoshoots with their hairdressers and makeup artists. "*

This is how man pays a heavy price for being in rivalry with avatars that resemble him there take me. He is mistaken for a CGI being, is guilty of being human and must agree to soon not more work to leave room for a simulation.

14 / What is the nature of the simulation?

14.1 / Video games timeline

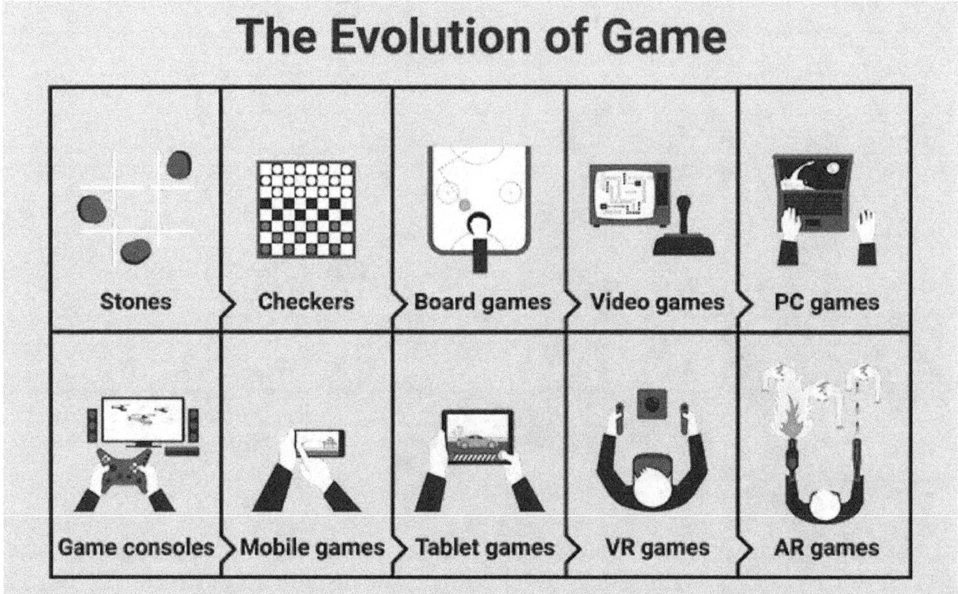

• **1940:** Edward U. Condon designs a computer that performs the traditional game Nim in which the players try to avoid picking up the last matches.

• **1947:** Thomas T. Goldsmith Jr. and Estle Ray Mann file a patent for a device cathode ray tube entertainment. Their game puts the players challenged to shoot a weapon at a target.

• **1950:** Claude Shannon and Alan Turing create chess programs.

• **1952:** AS Douglass creates *OXO Noughts and sticks* in the UK and *Tic-tac-toe* in the US United, on the Cambridge EDSAC computer in as part of his research on interactions man-machine.

• **1954:** The programmers of Los Laboratories Alamos from New Mexico develop the first blackjack program on a computer IBM-701.

• **1955:** The American army designs Hutspiel, in which red and blue players (NATO and Soviets) are at war.
• **1956:** Arthur Samuel presents his program of computer auditors, written on an IBM-701, at national television. In 1962 the program beats a master's degree in checkers.
• **1957:** Alex Bernstein writes the first program of chess advanced enough to assess four half-strokes forward on an IBM-704 computer.
• **1958:** Willy Higinbotham creates a tennis game on an oscilloscope and an analog computer for Brookhaven National Laboratory. He anticipated later video games like Pong.
• **1959:** MIT students create Mouse in the Maze on the TX-0 computer at MIT. The players draw a maze with a light pen, then a mouse walks through the labyrinth at the cheese search.
• **1960:** Computer programmer John Burgeson develops a baseball simulation computer science. A month later he performs this program on an IBM 1620 computer.
• **1961:** The Raytheon Company develops a computer simulation of the world conflict of cold war for chiefs of staff Americans. Although sophisticated, the simulation turns out to be too complex. Raytheon creates then a more accessible analog version called Grand Strategy.
• **1962:** Steve Russell, MIT student, invents *Spacewar!* , the first video game on computer. The game is spreading across the country.
• **1963:** The US Department of Defense complete a war game known as by *STAGE* (Simulation of Total Atomic Global Exchange) which shows that the United States would defeat the Soviet Union in a war nuclear.
• **1964:** John Kemeny creates the sharing system computer time and the language of BASIC programming in Dartmouth. Both allow students to easily write games computer science. Soon many games are created.
• **1965:** After Dartmouth defeated Princeton 28-14 in football, a student from Dartmouth program the first football match by computer.

• **1966:** While waiting for a colleague, Ralph Baer conceives the idea of playing a video game on television and writes down his ideas which become the basis of his development of television video games.

• **1967:** Ralph Baer develops his *Brown Box* , the video game prototype that allows users to play tennis and other games.

• **1970:** Scientific American publishes the rules of *LIFE* in the "Mathematical Games" section of Martin Gardner. In this simulation, cells isolated or overcrowded die, while others live and reproduce.

• **1971:** Minnesota students Don Rawitsch, Bill Heinemann and Paul Dillenberger create *Oregon Trail* .

• **1972:** Nolan Bushnell and Al Alcorn d'Atari are developing an arcade table tennis game. *Pong* was born.

• **1974:** Two decades before Doom, Maze Wars presents the first person shooter by taking the players through a labyrinth of passages made from wire graphics.

• **1977:** Atari launches the video computer system, more commonly known as the *Atari 2600* . Equipped with a joystick, interchangeable cartridges, games color and switches to select games and set the difficulty levels, it makes millions of Americans video game players at the House.

• **1978:** Taito's *Space Invaders* arrives in Japan, causing a shortage of 100 yen coins.

• **1979:** The toy manufacturer Mattel completes its portable electronic games with a new console, Intellivision. Better graphics and more sophisticated controls than the Atari 2600, Mattel sells three million units.

• **1980:** A missing slice of pizza inspires Toru Namco's Iwatani to create Pac-Man, which will be put up for sale in July 1980. That year, a version of Pac-Man for Atari 2600 becomes the first arcade hit to appear on a living room.

• **1981:** Jumpman. better known as Mario is created by Shigeru Miyamoto, making him the star of a later Nintendo game.

• **1982:** Disney draws on the craze for video games by releasing the movie *Tron* . An arcade game also becomes a success.

- **1983:** The multiplayer game takes a huge step forward in front with Dan Bunten's *MULE* .
- **1984:** Russian mathematician Alexey Pajitnov creates *Tetris* , a simple yet very addictive. Five years later, Nintendo integrated it with every new Game Boy.
- **1987:** Shigeru Miyamoto creates *Legend of Zelda* , SSI wins video game license for *Dungeons and Dragons* and *Sierra's Leisure* .
- **1988:** John Madden Football introduces realism grid in computer games.
- **1990:** Microsoft offers a video game version of the classic card game with Windows 3.0. *Solitary* becomes one of the most popular electronic games popular of all time and provides games easy-to-play *casuals* like *Bejeweled* .
- **1991:** Sega finds its iconic hero in *Sonic the Hedgehog* .
- **1992:** *Dune II* from Westwood Studios establishes the popularity of real-time strategy games.
- **1994:** Blizzard launches *Warcraft: Orcs and Humans*, a real-time strategy game that introduces million players in the world of Azeroth.
- **1995:** Sony released the PlayStation in the United States, and PlayStation 2 in 2000, becoming the console dominant domestic. Sega will leave the sector home consoles a little later.
- **1996:** *Lara Croft* makes her star debut from Eidos *Tomb Raider* adventure game .
- **1997:** The machine triumphs over man while IBM's supercomputer chess program *Deep Blue* defeats world champion Gary Kasparov in a match.
- **1998:** *Legend of Zelda: Ocarina of Time* transports players in the world of Hyrule.
- **2000:** The *Sims* of Will Wright are characters representing "real life". This is not the first simulation game - *Utopia* on Intellivision (1982), *Populous* by Peter Molyneaux (1989), *Sid Meier's Civilization* (1991) and *SimCity* (1989) preceded it.
- **2001:** Microsoft enters the video games market with Xbox and games like *Halo: Fighting Evolved* .

• **2002:** The American army launches the video game of
the U.S. military to help recruit and communicate with a new
generation of players. The Woodrow Wilson International Center
for Scholars launches the Serious Games Initiative for
encourage the development of games that address
policy and management issues.
• **2004:** Nintendo maintains its dominance over the
handheld market with Nintendo DS.
• **2005:** Microsoft's Xbox 360 brings high definition realism to the
gaming market.
• **2006:** The Nintendo Wii allows players to leave the sofa and
move thanks to innovative remote controls sensitive to
movement.
• **2007:** Take your instrument and play *Rock Band* .
• **2011:** *Skylanders: Spyro's Adventure* becomes the
first hit of augmented reality. Two more years *Disney
Infinity* later joins the ranks of hybrids toys-video games.
• **2013:** *Gone Home, The Last of Us* , *and Papers,
Please* usher in a new wave of video games mature who
confront players with choices difficult emotional worlds
ethically complex.
• **2014:** Free-to-play becomes a model dominant commercial
while blockbusters such as *CrossFire* , *Clash of Clans* , *World of
Tanks* and even *Kim Kardashian: Hollywood* have sales of
hundreds of million dollars through payments by
microtransaction.
• **2016:** Players are looking for Pikachu and Horsea in the real
world with free *Pokémon Go* from Niantic.
• **2017:** Nintendo's Switch introduces the first mobile hybrid video
game console with games as *Legend of Zelda: Breath of the
Wild* and *Super Mario Odyssey* .
• **2019:** Millions of players log on to watch a virtual asteroid
destroy the game map massively popular online battle royale
from Epic Games, *Fortnite* . This game grossed 2.4
billion dollars in 2018.

14.2 / NPC or RPG

The current versions of the simulation hypothesis are
mostly based on our recent progress in video game
technology. Oxford Professor Nick Bostrom popularized the idea
in his article in 2003, *"Do you live in a simulation?"* '
Since then, Elon Musk, among many others, has defended
this argument, which claims that if a civilization can day arrive at
the point of simulation, the ability technology to create such a
realistic virtual world than the physical world, then this has
already happened. There is probably has many more simulated
beings in the virtual worlds than real beings in the reality of
based. Since there is no easy way to tell if we sums of simulated
beings or not, we use numbers and they suggest we are more
likely to be in a simulation than the opposite. This is what has
leads Elon Musk to assume that the probability that
we are not in a simulation is one in Billions. This concept is
somewhat the equivalent modern idea dating back thousands of
years.
Buddhism and Hinduism largely emphasize before the idea that
we live in an illusory world called *maya*. Today, a large number of

researchers, including the Nobel Prize-winning physicist
Stanford's George Smoot and Leonard Susskind think that
physics already shows us through effects quantum that we are in
a simulation.
Sam Altman of Y Combinator said he knew several billionaires
who were financing their own means of exiting the simulation.
There are two versions of the simulation hypothesis which
can be best understood through the lens of games
video.

1. The NPC version : All entities within the simulation (we) are in
fact characters not players (NPCs) in video games - characters
simulated artificial devices that do not exist outside of the game.

2. The RPG version : This is the model suggested by religions of
the world, we exist as entities conscious outside the simulation
and let's play simply a role or assume an avatar, an incarnation,
in a simulated world.
If we are NPCs in one game only, there there may be restrictions
(code) that prevent us to understand that we are inside a game
video. As we have seen with people who do the experience of
taking LSD, the `` altered " consciousness maybe can decode
the simulation. Remember those who perceive grids in the sky as
if they were the limits of the world. In the NPC scenario,
if we found out that we are in a simulation and that the simulators
did not want we knew they could just rewind the
simulation, erasing those memories, instead of stopping
everything.
It might have happened before, but the man doesn't
would just not remember. Philip K. Dick thought
that it had already happened several times and wrote his
novel, *The Man In the High Castle,* because it claimed
remember an alternate timeline where the Nazis had
won World War II, a timeline which was rewound by the
simulators. Maybe he remembered from his previous lives.

If we are RPG characters in a simulation, this means that most of us already exist as conscious entities outside of the simulation. But we ignore this fact because the simulation is indistinguishable from reality. A bit like in the film as in `` *Matrix* '' where for Neo, stop the simulation would mean waking up from the illusion.

14.3 / The simulation and its creator

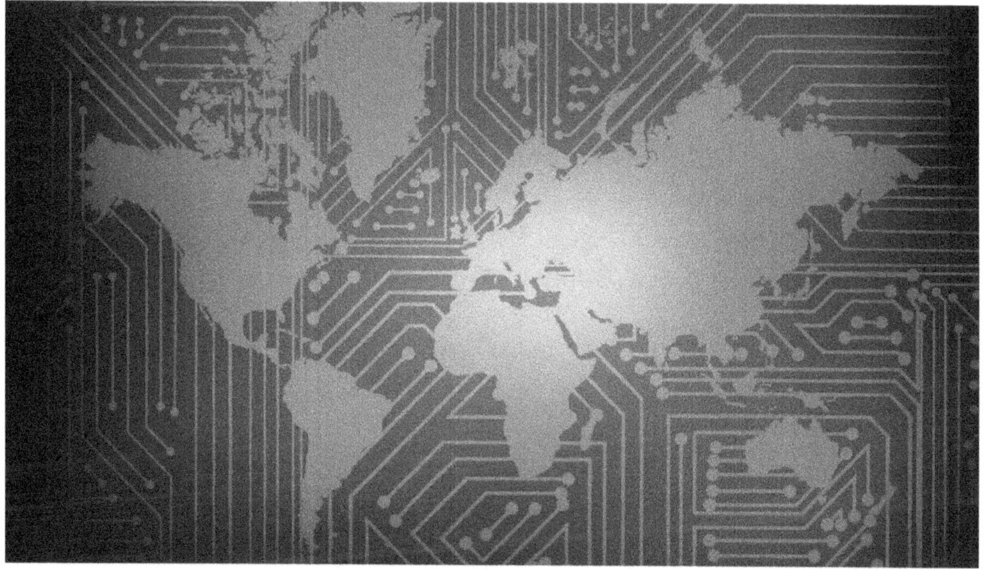

Our world can be simulated by different types of creators:

• **God:** transcendent (beyond our capacity to be able to understand) and / or mathematics
• **Humans:** human brain in a state of dream, of narration
• **Super intelligent AI:** young AI / quantum computer
• **Boltzmann's brain:** suggests that it is more likely for a brain to form spontaneously in a void (with a false memory of having existed) that so that our universe was created in the way that science explains it to us
• Aliens : Alien AI

Nature can be:

• Computer / holographic type
• Bio holographic (Biological Earth / Space holographic)
• Biological and then immersed in " others worlds " or dimensions (nested spheres)

The most likely type of simulation is:

• Centered on the observer
• Cheap
• Scientific or entertaining simulation of a possible world history close to singularity
• A simulation specially designed to obtain a tailor-made personal benefit, such as a resurrection simulation

The infinite simulation

We could be in the following channel:

1. Humans live in a simulation created by the species A.
2. Species A lives in a simulation created by species B.
3. Species B lives in a simulation created by the species C.
4. Etc.

Philosophical arguments for the existence of a divine creator or not

Nick bostrom

Let us recall here the philosopher Nick Bostrom who postulates that if a civilization gets to the point of creating a simulation HD, so she can create billions of civilizations simulated with billions of beings, because everything she has need more computing power.
If we are conscious beings, we are more likely to be a simulated being than a being organic.

Morality

1. If the creator does not exist, objective morality does not exist.
2. An objective morality exists.
3. Therefore, the creator exists.

Contingency

1. Everything has a reason why it exists - either by necessity of its own nature, either because it was caused by something else.
2. If the universe has a reason why it exists, it is that the creator made it exist.
3. The universe exists.
4. Therefore, the Creator made the universe exist.
5. Therefore, the creator exists.

Mathematical design

1. If the creator does not exist, the applicability of mathematics to the physical world is just one coincidence.
2. The applicability of mathematics to the world physics is not just a coincidence.
3. Therefore, the creator exists.

Ontology

1. What nothing equal or greater can be conceived (the creator) may exist. (Premise)
2. Suppose the creator may be missing to exist. (Assumption)
3. A being who cannot fail to exist is more great than a being who can fail. (Premise)
4. If what one cannot conceive neither equal nor superior cannot fail to exist, so it is not what one cannot conceive neither equal nor higher (after 3)).
5. But this is a contradiction.
6. Therefore, what nothing equal or more great cannot be conceived, therefore a creator, may fail to exist.

If the simulation stopped

Nick Bostrom listed the risk of stopping the simulation as one of the existential risks. As short simulations are more likely than long universe simulations, the end of the simulation could be close.
Here is why the simulation can be interrupted, from more likely less likely depending on understanding scientists :
1) Overload of the calculation resources of the reality of based.
The IT resources needed to run simulation of humanity could exceed the available resources.
2) The singularity simulation ends shortly after the creation of the AGI because because the super intelligent AI of the newborn will probably create his own.

3) Simulation of global catastrophic risks that could test how the human race evolves in the bipolar world and how often this results in a nuclear disaster, which civilization cannot recover.
4) Bugs or viruses. This could appear at a level where the simulation must be reloaded.
5) Awareness of simulation. The beings inside the simulation begin to consciously realize or not that they are in a simulation, it is no longer a simulation and cannot be used as such.
6) End of the game. simulation finally solves the task unknown to mankind, it is no longer necessary run the simulation.
7) Accidental termination. some world events real end the simulation. a person could be wake up from a dream if his alarm rings or a computer could suffer a power failure.
8) Accumulation of simulation endings nested. higher level simulation (closer of real reality) in nested simulations goes out.
9) Natural ending.

14.4 / After physical death

Religions and the soul after death

According to Buddhism, the soul is divided into six destinies with the consequences of his previous acts and the possibility of rebirth. For Christianity, the soul is eternal and according to his behavior in his life terrestrial, observed by recording angels, we we are heading either to hell or to paradise. In Hinduism, there is the notion of retribution for faults and acts of kindness in his incarnation. Man inherits of his previous life and his good or bad deeds and can be reincarnated as an animal or a plant. According to Islam, the soul is detached from the body after death and survives until the day of judgment when we go to hell or in paradise according to his past acts. For the Judaism, death is not the end of life, the soul returns to its creator as the body turns to dust. The teachings speak of a resurrection occurring at the end of time inaugurated by the coming of the messiah. The death is therefore only a stage of life, the soul joining the soul ancestors. Finally, according to esotericism, the body etheric detaches itself from the physical body and remains linked to the astral body several days. The separated astral body is in hell the time to repair his faults, and therefore improve the karma for future reincarnations. Forgetting lives previous ones occurs during the new reincarnation.

After death; eternal nothingness?

Some people think that after death all conscious experience ceases. The thought that this life is the only thing that exists and that after it comes a total and eternal darkness is very distressing.
Finally, what exactly is 'nothing'? A hole in a doughnut is there something or nothing? A similar logic applies to shadows; are they something or nothing? Even though the holes and the shadows are nothing, they seem to have something to do with thing. It is impossible to imagine the doughnut hole without the doughnut itself. Likewise, that it is impossible to imagine a shadow without the object and the light that plans.
Maybe what exists after death is just as much related to the existence that shadows are to an object. This means that even dead we are not totally separate from existence and from this world. In this, death does not mean that we no longer exist, but that we transform ourselves.

Immortality in simulations

The existence of a very large number of simulations created implies that there are other copies of each person in different simulations. This ties in with the theory of parallel universes. There are a very large number but finished number of people possible, limited by combinations atoms, and if the number of simulations is greater than number of possible people, people will repeat themselves in the simulations. If today's

humanity is in an ancestral simulation, it will eventually be
performed several times with small variations as well
many of its elements will be repeated exactly or with small
variations, including humans.
Thus, if its simulation was disabled, other simulations with other
copies would continue to exist; it is a form of immortality.

Spirits / Ghosts

Spirits are souls without a bodily envelop which maintain an
ethereal body. Spiritism think that the spirit survives and that is
what allows communicate with the spirit of the dead. They would
be around of us and are most often with people
that they loved while they were alive. The perception of
Spirits being highly developed, they can know the
deep thoughts and desires.
According to computer scientist Curry Guinn of the University of
North Carolina at Wilmington, reports phenomena such as
ghosts, déjà vu and strange coincidences could in fact be
problems in the matrix. These problems could turn out to be
another proof that humanity does not live part in a real
universe. According to Guinn, we must pay attention and check
for developing bugs.
Deja vu, as in the movie *Matrix* when Neo sees a cat in the same
place several times, maybe some kind of bug.
*"Ghosts, ESPs, and coincidences can be problems. The laws of
physics in our universe seem designed with a set of constants
that make life based on carbon. Where are the edges? "*
Is there any ghost data left in a universe holographic?
Assuming we're the equivalent of a Sim's in a simulation, what
happens to an avatar when it dies? Sometimes, by oddity,
because of a glitch or bug, AI-led creatures respond as if
they were alive for them but absent / invisible for most other
creatures. In fact, some could however see a ghost digital. Now
consider a game where the developers would not have expected
that something as it could happen, so why not in a more

sophisticated version of it in a computer as big as our universe? When deleting a file on our computer, it ends up in the recycle bin, pending actual deletion. Would it be possible while ghosts are data almost between two states: Suppressed but available in the trash? A kind of entanglement!

Reincarnation; a mind-blowing case

James Madison Leininger was born on April 10, 1998 in San Francisco, he is the son of Bruce and Andrea Leininger. Expressions from the memory of life little James' past manifested themselves between the age of two and five years after moving to Lafayette. The combination of his detailed memories and the ability to of parents to verify them makes a very relevant case.

When James was 2, his mother noticed a basket filled with toys and boats and took a small plane to helix to give it to James, adding:

"Look, there's even a bomb below!" James replied, *'It's not a bomb, mom. It's a tank".* Talking about it with her husband, she learned later than a drop tank is a tank of additional fuel installed on an aircraft to extend its reach. But how could James have know that at his age? Then the child began to do very frequent nightmares in which he cried often: ' ' *Plane crash! Plane on fire! The little man cannot go out! "* …

James later told his parents that the little one man was himself and that his plane had been shot down by the Japanese. About two weeks later he added more details: his name was James and he had piloted a Corsair that came from the Natoma. During the three following months, James explained that he had had a friend, a fellow pilot named Jack Larsen, and that he had been shot dead near Iwo Jima.
Bruce Leininger, James' dad, uncomfortable with the idea of the reincarnation of his son, began to Internet research. He thus discovered that the USS Natoma Bay was an aircraft carrier, having served in the Pacific during World War II, was part of an Iwo Jima operation, and that a named pilot Jack Larsen was based on the ship. Bruce approached Natoma Bay veterans, including Larsen.

James Huston, Jr.   James Leininger

Attention then turned to James McReady Huston Jr. who was killed near Iwo Jima at the age of 21 years. Little James' statements seemed match. One exception was that Huston had was killed in an FM2 Wildcat, not a Corsair. However, a visit to Huston's sister, Anne Barron, revealed a photograph of the latter standing in front of a Corsair, proving that he had already flown this plane. Then testimonies confirmed that the plane of Huston had exploded before crashing, confirming the James story. Anne Barron also checked other details that James had given on his previous family, y understood the problems caused by the alcoholism of his dad. After speaking with James, she was convinced that he was indeed his reincarnated brother. James explained also that he remembered choosing Bruce and Andrea for parents, and gave some details on the period preceding its conception. When his parents gave him asked why he had named his three dolls GI Joe Billy, Walter and Leon, he replied that it was because that he had met them when he arrived in paradise. The parents later learned that three companions squadron of Huston having been killed before him were called Billy Peeler. Walter Devlin and Leon Conner, like its GI Joe.

After the release of *Soul Survivor* in 2009, Fox 8 News broadcast a documentary featuring interviews with Natoma Bay veteran Leo Pyatt describing how little James recognized the other veterans, and Huston's sister, Anne Barron. The report tells also how a Japanese TV channel has brought the family to Iwo Jima where Huston died, and where James was able to pinpoint the exact location. In a 2013 clip on Fox and Friends, the then 15-year-old James, described how the nightmares ceased after a spiritual liberation experienced at the site of the accident. James went on to suggest that reincarnation may be the source of what appears to be innate knowledge. He added that he sometimes remembers his previous life, but that he is moving forward in his current life.

Science's perspective on the afterlife

The near-death experience NDE has been described in 1975 by Raymond Moody in " *Life after death* " after having gathered the testimony of several people with NDE. Every time the same words come back: body seen from above, perception of a tunnel, attraction to a being of light, often one close to the nobody. These experiences have always been rejected by scientists thinking of a hallucination due to retinal activity suffering from lack of oxygen or a trauma. Yet a study published in October 2014 by scientists from the University of Southampton changed everything. While they were in condition of clinical death, 40% of patients spoke of a feeling of consciousness and spoke of an exit from the body during which they were able to observe scenes that actually took place. Consciousness seems therefore continue after death.

15 / Towards a theory of everything? (Epilogue)

Could relativity and quantum physics agree?
Andrzej Dragan of the Department of Physics of Warsaw
University and Artur Ekert University of Oxford believe in it and
propose their theory in a recent article published by the New
Journal of Physics.
Even if their work is for the moment purely theoretical, it opens
up very interesting perspectives to understand what exactly
could have happened at the appearance of the universe or what
we could find in
a possible black hole. The two researchers have shown
that at the mathematical level it is possible to deduce the
relativity non-locality and superposition effect restraint. It would
follow that it is feasible to observe phenomena moving so much
at the speed of light than above. Relativity postulated that no
object could not move at the speed of light and at above. In their
hypothesis, we find, in the field of relativity, the quantum
phenomena of non-locality or superposition.
Quantum mechanics have been waits for a deeper theory to
explain nature of its mysterious phenomena.
Could it be that a global theory, integrating physics
relativistic and quantum physics, emerges this new decade ?
In the age of the internet, sophisticated software and CERN,
modern scientists have no more excuses for ignore the reality of
our paradigm.

16 / References

1 / Introduction

https://home.cern/science/physics/standard-model
https://www.letemps.ch/sciences/physiciens-apportent-a-definitive-mass-neutrino-proof

2 / The big bang; a pulse

https://www.godandscience.org/apologetics/designun.html

https://www.forbes.com/sites/fernandezelizabeth/ 2020/08/18 / universe-may-have-started-in-a-big-bounce-rather-than-a-big-bang-scientists-say / # 45a540607202

3 / Abiogenesis is impossible

http://edusofad.com/www/demo/wged-scp/demo/saw1m01e1.php

https://www.mun.ca/biology/scarr/4270_Redi_experiment.html

4 / The atom is 99% empty

https://phys.org/news/2017-02-atoms-space-solid.html

https://www.forbes.com/sites/startswithabang/ 2020/04/16 / you-are-not-mostly-empty-space / # 6e0692002c2b

https://www.futura-sciences.com/sciences/questions-answers / physics-if-atoms-are-composite-empty-matter-is-it-not-transparent-13256 /

5 / Nothing goes faster than light

https://ploum.net/pourquoi-ne-peut-on-pas-depasser-la-speed of light/
*https://gizmodo.com/5-reasons-our-universe-might-*actually-be-a-virtual-real-1665353513

http://www.courselectricite.com/electromagnetisme.html
https://fr.wikipedia.org/wiki/%C3%89ther_(physique

6 / The Goldilocks area

https://www.abc.net.au/news/science/2016-02-22/goldilocks-zones-habitable-zone-astrobiology-exoplanets / 6907836

https://laterreestconcave.home.blog/2020/05/24/non-la-nasa-did-not-find-a-parallel-universe-in-antarctica-the-land-is-open-to-poles /

https://laterreestconcave.home.blog/2020/05/29/terre-hollow-vs-concave-Earth-or-the-sf-face-to-reality /

https://laterreestconcave.home.blog/2019/01/19/les-polar-lights-proof-of-the-Earth-concave-and-of-opening-of-poles /

https://fr.wikipedia.org/wiki/Th% C3% A9ories_de_la_Terre_creuse

https://ncse.ngo/inside-out-and-round-about-part-2

https://cyprustar.wordpress.com/2017/12/28/a-geocosmos-mapping-outer-space-into-a-hollow-Earth-mostafa-a-abdelkader /

https://www.globalgreyebooks.com/read-online/cellular-cosmogony / read-online.html https://laterreestconcave.home.blog/2019/11/22/la-terre-concave-seen-by-professor-paolo-emilio-amico-roxas /

https://laterreestconcave.home.blog/2019/08/13/le-solar-system-is-geo-heliocentric /

https://cyprustar.wordpress.com/2020/05/24/non-la-nasa-did-not-find-a-parallel-universe-in-antarctica-la-
land-is-open-to-poles /

https://laterreestconcave.home.blog/2019/07/27/espace-time-in-the-Earth-a-4-dimensions /

https://laterreestconcave.home.blog/2019/01/19/loctaedre-highlighted-by-the-radar-cmor /

https://laterreestconcave.home.blog/2019/01/19/le-vrai-model-of-our-Earth /

https://laterreestconcave.home.blog/2019/01/19/discovery-the-golden-number-found-in-the-dimensions-de-la-terre-concave /

https://www.franceculture.fr/sciences/recherche-seti-pas-one-whisper-of-alien-tech

https://cyprustar.wordpress.com/2020/05/24/non-la-nasa-did-not-find-a-parallel-universe-in-antarctica-la-
land-is-open-to-poles /

https://www.courrierinternational.com/article/cosmologie-evidence-of-a-parallel-universe-discovered-in-
Antarctic

7 / intelligent design

http://vedicsciences.net/articles/intelligent-design.html

http://www.db-gersite.com/HISTOLOGIE/EPITHDIG/ intestine / intes2 / intes2.htm

https://www.secretsinplainsight.com/hidden-universal-symmetry /

https://theinfiniteyou.info/the-intelligent-design-found-throughout-nature /
https://people.howstuffworks.com/intelligent-design.htm

https://www.josh.org/what-is-the-best-evidence-for-intelligent-design-interview-with-brian-johnson /

https://www.researchgate.net/figure/De-la-structure-primer-with-quaternary-structure-of-proteins-Dapres_fig2_43642467

https://www.researchgate.net/figure/De-la-structure-primer-with-quaternary-structure-of-proteins-Dapres_fig2_43642467

https://reasons.org/explore/blogs/the-cells-design/read/ the-cells-design / 2019/05/29 / biochemical-grammar-communicates-the-case-for-creation

https://reasons.org/explore/blogs/the-cells-design/read/ the-cells-design / 2019/05/29 / biochemical-grammar-communicates-the-case-for-creation

https://biologiedelapeau.fr/spip.php?article9

https://www.treknature.com/gallery/photo256103.htm

https://www.creationest.com/skin-intelligent-design.html

https://www.conservapedia.com/

General_and_Special_Evidence_for_Intelligent_Design_in_Biology_#_Positive_Evidence_for_Design

https://outlookmag.org/the-human-eye-a-case-for-intelligent-design /

https://ndgperception.weebly.com/i-le-fonctonnement-de-loeil.html

https://www.treehugger.com/how-golden-ratio-manifests-nature-4869736

https://indianexpress.com/article/explained/the-mysterious-golden-ratio-why-is-it-everywhere-now-in-human-skull-6067588 /

8 / The shape of the universe

https://www.nature.com/articles/s41550-019-0906-9
https://www.loop-doctor.nl/wp-content/uploads/2019/10/Loop-Doctor-Investigates-Why-the-universe-is-spherical_V2.pdf

9 / The holographic universe

https://www.institutneuroperformance.com/en/neurotherapy-quantum-physics-and-universe-holographic /

https://arxiv.org/abs/1901.10489

https://cosmosmagazine.com/physics/universe-shouldn-t-exist-cern-physicists-conclude

https://www.sciencedirect.com/science/article/pii/S1571064513001188

http://www.linternaute.com/science/espace/dossiers/06/

theory-of-everything / 8.shtml

*https://www.sciencedaily.com/releases/
2019/06 / 190619103151.htm*

*https://asgardia.space/en/news/Confirmed-There-Is-No-
Objective-Reality-According-to-Quantum-Mechanics*

*https://www.gizhub.com/physicist-discovers-reality-just-
may-computer-simulation*

*https://www.nature.com/news/simulations-back-up-
theory-that-universe-is-a-hologram-1.14328*

*https://www.quantamagazine.org/how-space-and-time-
could-be-a-quantum-error-correcting-code-20190103 /*

*https://trustmyscience.com/experience-quantique-
confirms-objective-reality-does-not-exist*

*https://futuristech.info/posts/opinion-we-are-the-sims-
living-life-in-this-mega-computer-simulation*

*https://hackernoon.com/are-we-already-in-the-matrix-
7492e89be433*

**10 / The fascinating similarity between the neural network
and dark matter**

*https://www.webastro.net/forums/topic/55206-mati
% C3% A8re-black-and-neurons /*

11 / Consciousness and the brain

http://www.neuromedia.ca/le-systeme-limbique/

https://www.kurzweilai.net/where-is-self-awareness-

located-in-the-brain

https://royalsocietypublishing.org/doi/10.1098/rsta.1998.0254

https://forum.tutoweb.org/topic/40390-sulcus-limitans/

https://en.wikipedia.org/wiki/Consciousness

https://www.science-et-vie.com/archives/qu-est-ce-que-consciousness-34832

https://medicalxpress.com/news/2017-04-evidence-higher-state-consciousness.html

https://openyourreality.com/dreams-explained-in-a-simulated-universe /

https://mydreamsymbolism.com/dreams-about-tsunamis-meaning-and-interpretation /

https://medium.com/@neurokinetikz/the-holographic-brain-e51b7185e677

https://www.reddit.com/r/LSD/comments/9n8voh/anyone_else_seen_the_grid_whilst_on_l /

https://www.reddit.com/r/Glitch_in_the_Matrix/comments / 80fbhw / weird_grid_in_the_sky /

https://www.lemonde.fr/blog/realitesbiomedicales/tag/psilocin /

https://dailygeekshow.com/analyse-reves-etude-realite/

https://www.cia.gov/library/readingroom/docs/CIA-RDP96-00788R001700210016-5.pdf

https://cyprustar.wordpress.com/2018/09/19/un-document-de-la-cia-suggests-that-we-live-in-a-hologram /

12 / Robots are aware of themselves

https://www.parismatch.com/Vivre/High-Tech/Notre-interview-of-Ai-Da-the-first-robot-artist-1634876
https://hitek.fr/actualite/robot-pris-conscience-de-lui-meme_6666

13 / Distinguish between reality and fiction

https://www.virtualhumans.org/article/can-virtual-influencers-have-real-influence

https://cyprustar.wordpress.com/2020/10/20/les-virtual-influencers-steal-the-job-from-real-people /
https://fr.wikipedia.org/wiki/Fiction#Fiction_et_r % C3% A9alit% C3% A9

14 / What is the nature of the simulation?

https://productivityhub.org/2019/09/01/we-need-to-find-out-if-we-are-living-in-a-simulation /

https://arxiv.org/ftp/arxiv/papers/1905/1905.05792.pdf

https://store.steampowered.com/app/1130560/ RPG_NPC_Simulator_VR /

https://kinesiologielimousin.com/aura-therapie-2

https://www.quebechebdo.com/dossiers/155491/quelle-is-the-vision-of-death-according-to-the-great-religions /

https://en.wikipedia.org/wiki/Boltzmann_brain
https://cyprustar.wordpress.com/2020/09/18/le-temps-tel-an infinite-match /

https://hastyreader.com/what-happens-when-you-die/

https://psi-encyclopedia.spr.ac.uk/articles/james-leininger # footnote13_0ji2wu0

https://www.reincarnationresearch.com/past-life-story-of-james-huston-jr-james-leininger /

https://curiosmos.com/our-universe-isnt-real-scientists-say-ghosts-could-be-signs-of-a-simulated-universe /

https://supernaturalmagazine.com/articles/are-ghosts-just-people-stuck-in-the-recycle-bin-of-the-holographic-universe

https://www.museumofplay.org/about/icheg/video-game-history / timeline

https://blog.treasuredata.com/blog/2019/04/18/6-gaming-trends-to-watch-now-get-ready-for-a-revolution /

https://www.si.edu/object/tv-game-unit-8-1968: nmah_1302003

15 / Towards a theory of everything? (Epilogue)

https://iopscience.iop.org/article/10.1088/1367-2630/ab76f7 / pdf

Author contact: cyprustar@gmail.com